Mathematical Research · Mathematische Forschung

Wissenschaftliche Beiträge
herausgegeben von der
Akademie der Wissenschaften der DDR
Institut für Mathematik

Band 22

Singular Ordinary Differential Operators
and Pseudodifferential Equations

by
J. Elschner

Singular Ordinary Differential Operators and Pseudodifferential Equations

by

Johannes Elschner

Akademie-Verlag · Berlin 1985

Verfasser
Dr. sc.nat. Johannes Elschner
Institut für Mathematik der
Akademie der Wissenschaften der DDR
1086 Berlin
Mohrenstraße 39

Die Titel dieser Schriftenreihe werden vom Originalmanuskript
der Autoren reproduziert

ISSN 0138 - 3019

Erschienen im Akademie-Verlag, DDR-1086 Berlin, Leipziger Straße 3-4

Lizenznummer: 202 · 100/408/85
Printed in the German Democratic Republic
Gesamtherstellung: VEB Kongreß- und Werbedruck, 9273 Oberlungwitz
Lektor: Dr. Reinhard Höppner
LSV 1035
Bestellnummer 763 388 4 (2182/22)
02600

INTRODUCTION

Various problems in physics and engineering lead to a linear ordinary differential equation whose coefficient of the highest derivative vanishes at certain points. Such an equation is called degenerate or singular, and a zero of the leading coefficient is said to be a singular point or singularity of the corresponding differential operator. In Chapters 1 and 2 of these notes we consider linear ordinary differential operators with a singular point at the origin:

$$A = x^q D^l + \sum_{0 \leq i < l} a_i(x) D^i \ , \ D = d/dx \ , \ q \in \mathbb{N} \ . \tag{1}$$

In case of the homogeneous equation $Ay = 0$ with analytic coefficients a_i, the investigation of singular differential equations has a long history. In particular, starting with the works of Fuchs and Poincaré in the second half of the last century, the asymptotic behaviour as $x \to 0$ of solutions to such equations has been studied by many authors; see e.g. Sternberg [1], Wasow [1], Ince [1]. In contrast to that, general results on the solvability of the degenerate inhomogeneous equation $Ay = f$ have been obtained only during the last fifteen years, using the methods of linear functional analysis.

In order to describe such a result and the material of these notes, it is necessary to recall some definitions. Let X and Y be linear topological spaces, and $A : X \to Y$ a linear operator with the domain of definition $D(A) \subset X$. If A is a continuous map of X into Y and $D(A) = X$, we shall write $A \in L(X,Y)$ and $A \in L(X)$ when $X = Y$. A is called normally solvable if its range $\operatorname{im} A = A(D(A))$ is closed in Y. The dimension $\dim \ker A$ of the kernel $\ker A = \{x \in D(A) : Ax = 0\}$ will be called the kernel index or nullity of A, and the deficiency $\dim Y/\operatorname{im} A$ of $\operatorname{im} A$ in Y will be called the deficiency index of A. If A is normally solvable and its kernel and deficiency indices are both finite, we say that A is a Fredholm operator, and its index is defined by $\operatorname{ind} A = \dim \ker A - \dim Y/\operatorname{im} A$. Furthermore, A is called Φ_+- (resp. Φ_--)operator if it is normally solvable and $\dim \ker A < \infty$ (resp. $\dim Y/\operatorname{im} A < \infty$). For Fredholm and $\Phi_\pm$-operators and their basic properties, we refer to the exposition in Goldberg [1], Prößdorf [1], Przeworska-Rolewicz and Rolewicz [1].

Malgrange [1], Komatsu [1] and Korobejnik [1] independently of each other found a general index formula for the operator (1) in the space $\mathcal{H}(\Omega)$ of analytic functions in a domain $\Omega \subset \mathbb{C}$, $0 \in \Omega$, which we state here in the special case of a simply connected domain:

If $a_i \in \mathcal{H}(\Omega)$ $(i=0,\ldots,l-1)$, then $A \in L(\mathcal{H}(\Omega))$ is a Fredholm operator with index $l-q$.

Thus we observe that the index of A coincides with the index of the principal part $x^q D^l$ in $\mathcal{H}(\Omega)$. This result is not true, in general, if (1) acts in spaces of differentiable functions on an interval $[a,b] \subset \mathbb{R}$, $0 \in [a,b]$. Since the end of the sixties there appeared a lot of papers concerning the index and the solvability properties of the operator (1) in such spaces. The aim of Chapters 1 and 2 of this work is to fit most of those results into a general framework and thus to give a review of the current state of this field. Chapter 1 deals with the special case of a Fuchsian differential operator and also serves as an illustration of typical methods and results in the theory of singular ordinary differential equations. In Chapter 2 we give a general index formula for the differential operator (1) in the space of infinitely differentiable functions and in weighted L_p spaces on an interval. Furthermore, we study kernel, range, normal solvability in weighted Sobolev spaces and hypoellipticity of the degenerate operator (1) as well as its index in the space of distributions. A preliminary version of Chapters 1 and 2 appeared in Elschner and Silbermann [2].

In Chapter 3 these results will partly be generalized to differential operators with a finite number of singular points on compact or infinite intervals. In case of Fuchsian differential operators on a finite interval, the index formula is applied to study essential selfadjointness and spectrum of boundary value problems for those operators.

We remark that there is an increased interest in the development of the theory of singular ordinary differential operators in connection with the very active fields of solvability theory for degenerate partial differential equations (see e.g. Bolley, Camus and Helffer [1], [2], Baouendi and Goulaouic [1], Baouendi, Goulaouic and Lipkin [1], Helffer and Rodino [1], Višik and Grušin [1], Elschner and Lorenz [3], Lorenz [1]) and numerical analysis for degenerate ordinary differential equations (cf. e.g. de Hoog and Weiss [1], [2], Natterer [1], Elschner and Silbermann [1]).

Chapter 4 illustrates the application of singular ordinary differential operators in the theory of partial differential equations. Relying on certain results in Chapters 1 and 2, we study local solvability as well as normal solvability and index in Sobolev spaces for some examples of elliptic partial differential operators degenerating at one point. Many problems in this field are still open.

Various boundary value problems in mathematical physics and complex function theory lead to singular integro-differential, or more generally, to pseudodifferential equations on a closed curve. A pseudodifferential operator is called classical if its symbol has an asymptotic expansion as a sum of symbols which are positive homogeneous in the covariable of decreasing orders (cf. Chapter 5); it is called non-elliptic or degenerate if the principal term of the symbol vanishes at certain points on the cosphere bundle of the curve.

A major part of Chapter 5 is devoted to the author's recent results on the index and the Fredholm property of degenerate classical pseudodifferential operators on a closed contour, though, for the lack of space, we have not covered all of the material in full generality. As an application of these results, theorems on the index and on existence and uniqueness of solutions of the degenerate oblique derivative problem in the plane are given. Furthermore, in Chapter 5 we have included an almost self-contained introduction to classical pseudodifferential operators on a closed curve. For the general theory of pseudodifferential equations, the reader is referred to Šubin [1], Taylor [1] and Treves [1]. An exposition of the theory of degenerate one-dimensional singular integral equations which have been studied somewhat earlier can be found in Prößdorf [1] and Michlin and Prößdorf [1].

Chapter 6 deals with the Galerkin method using periodic splines as test and trial functions for the approximate solution of pseudodifferential equations on a closed contour. It demonstrates the interplay between certain a priori estimates for pseudodifferential operators, namely the Gårding and Melin inequalities, and convergence results for Galerkin's method with splines for strongly elliptic and degenerate equations. For an introduction to the theory of splines and finite element methods, we refer to Aubin [1], de Boor [1] and Strang and Fix [1]. The reader should consult the introduction and the section "comments and references" in each chapter for more information on the contents of these notes and further references.

Except for Chapter 4, the material is rather self-contained. The reader is assumed to be familiar with linear functional analysis (see e.g. Goldberg [1], Prößdorf [1]). In Chapter 4 some previous knowledge of elliptic differential operators on manifolds is desirable (cf. Narasimhan [1], Agranovič and Višik [1]).

Throughout the book the following notation is used. For a domain $\Omega \subset \mathbb{R}^n$, let $C^\infty(\Omega)$ ($C_0^\infty(\Omega)$) be the set of infinitely differentiable functions (with compact support) in Ω, and $C^\infty(\bar{\Omega})$ the set of all infinitely differentiable functions in Ω which together with all derivatives continuously extend to the closure $\bar{\Omega}$ of Ω. For $\Omega = (a,b) \subset \mathbb{R}$, we simply write $C^\infty(\Omega) = C^\infty(a,b)$, $C^\infty(\bar{\Omega}) = C^\infty[a,b]$ etc. The support of a function u is denoted by supp u. In $C^\infty(\bar{\Omega})$ (resp. $C_0^\infty(\Omega)$) one can introduce the topology of a Fréchet (resp. locally convex) space; see Hörmander [2], Robertson and Robertson [1]. The bilinear form

$$\langle u,v \rangle = \int_\Omega uv\, dx$$

on $C_0^\infty(\Omega) \times C_0^\infty(\Omega)$ extends to a duality between $C_0^\infty(\Omega)$ and the locally convex space $\mathcal{D}'(\Omega)$ of all distributions in Ω. For $A \in L(C_0^\infty(\Omega))$, the transpose ${}^tA \in L(\mathcal{D}'(\Omega))$ of A is defined by

$$\langle {}^tAu,v \rangle = \langle u,Av \rangle, \quad u \in \mathcal{D}', \ v \in C_0^\infty.$$

Finally, if M is an n-dimensional infinitely differentiable manifold with or without boundary, let $C^\infty(M)$ be the set of all infinitely differentiable functions on M; see Narasimhan [1]. Other notation is either standard or defined upon introduction.

Acknowledgements

My interest in degenerate differential and integral equations was stimulated by my teacher Prof. Dr. S. Prößdorf, and I thank him for helpful discussions and permanent attention in my scientific development. The results in Chapter 4 arose in cooperation with Dr. M. Lorenz, and I am grateful to him for going through the manuscript and making various suggestions and corrections. Further I want to express my gratitude to Dr. G. Schmidt who read parts of the manuscript. In particular, his remarks led to several improvements in Chapter 6. I would also like to thank Mrs. Ch. Huber for typing the present monograph. Finally, I want to acknowledge the Akademie-Verlag, especially Dr. R. Höppner, for effective cooperation and including this publication in the series "Mathematical Research".

Berlin, December 1983

J. Elschner

CONTENTS

1. FUCHSIAN DIFFERENTIAL OPERATORS WITH ONE SINGULAR POINT

In this chapter we consider the Fuchsian differential operator

$$(Ay)(x) = [x^l D^l + \sum_{0 \le i < l} a_i(x) x^i D^i] y(x), \quad Dy = y' = dy/dx , \qquad (1.0.1)$$

of order l which has a singular point at the origin. We have chosen this simple class of operators in order to introduce the methods which are used in the study of more general degenerate ordinary differential equations in Chap. 2. Furthermore, Fuchsian differential equations occur in several applications in mathematical physics and mechanics, which justifies a more detailed investigation of those operators.

Before studying the operator (1.0.1) in certain spaces of differentiable functions, we consider the corresponding Euler differential operator

$$A_0 = x^l D^l + \sum_{0 \le i < l} a_i(0) x^i D^i. \qquad (1.0.2)$$

It turns out that A and A_0 are simultaneously Fredholm operators in the corresponding spaces and their indices coincide since A is a small perturbation of A_0 in a certain sense. Thus A_0 can be considered as the principal part of A.

1.1. Spaces

In this section we define function spaces in which index and solvability properties of the operator (1.0.1) shall be investigated, and collect some of their properties.

1.1.1. Let $b > 0$ and $L_p(0,b)$, $1 \le p \le \infty$, be the space of all complex-valued measurable functions on $(0,b)$ for which the norm

$$\| y \|_{L_p(0,b)} = \left(\int_0^b |y(x)|^p dx\right)^{1/p} , \quad 1 \le p < \infty ,$$

$$\| y \|_{L_\infty(0,b)} = \operatorname*{ess\,sup}_{0 \le x \le b} |y(x)|$$

is finite.* For $k \in \mathbb{N}$, $W_p^k(0,b)$ will denote the Sobolev space of all functions y for which $y^{(k-1)} = D^{k-1} y$ exists and is absolutely continuous on $[0,b]$ and the norm given by

$$\|y\|_{W_p^k(0,b)} = \sum_{0\le i\le k} \|y^{(i)}\|_{L_p(0,b)}$$

is finite. We set $L_p = W_p^0$. Let $C^k[0,b]$, $k \in \mathbb{N}_0 = \mathbb{N} \cup \{0\}$, be the space of k times continuously differentiable functions on $[0,b]$ with norm

$$\|y\|_{C^k[0,b]} = \sum_{0\le i\le k} \max_{0\le x\le b} |y^{(i)}(x)| .$$

Furthermore, we introduce the spaces with weights

$$L_p^\varrho(0,b) = W_p^{0,\varrho}(0,b) = \{ x^\varrho y : y \in L_p(0,b)\}, \quad \varrho \in \mathbb{R},$$

$$C^{0,\varrho}[0,b] = \{ y \in C^0[0,b] : x^{-\varrho}[y(x)-y(0)] \in C^0[0,b]\}, \quad \varrho \ge 0,$$

and for $k \in \mathbb{N}$, $\varrho \ge 0$,

$$W_p^{k,\varrho}(0,b) = \{y \in W_p^{k-1}(0,b) : y^{(k)} \in L_p^\varrho(0,b)\},$$

$$C^{k,\varrho}[0,b] = \{ y \in C^{k-1}[0,b] : y^{(k)} \in C^{0,\varrho}[0,b]\} .$$

$W_p^{k,\varrho}$ and $C^{k,\varrho}$ are Banach spaces with norms defined by

$$\|y\|_{W_p^{k,\varrho}(0,b)} = \|y\|_{W_p^{k-1}(0,b)} + \|x^{-\varrho} y^{(k)}\|_{L_p(0,b)},$$

$$\|y\|_{C^{k,\varrho}[0,b]} = \|y\|_{C^{k-1}[0,b]} + \|x^{-\varrho}[y^{(k)}(x)-y^{(k)}(0)]\|_{C^0[0,b]} + |y^{(k)}(0)| ,$$

where the first term on the right-hand sides is omitted for k=0.

Finally, for $k \in \mathbb{N}_0$ and $\lambda \in (0,1]$, let $H^{k,\lambda}[0,b]$ be the space of all functions $y \in C^k[0,b]$ for which $y^{(k)}$ satisfies a Hölder condition with exponent λ on $[0,b]$, equipped with the norm

$$\|y\|_{H^{k,\lambda}[0,b]} = \|y\|_{C^k[0,b]} + \sup_{0\le x<t\le b} |y(x)-y(t)|\,|x-t|^{-\lambda} .$$

<u>1.1.2.</u> For $\varrho \ge 0$, we introduce the closed subspaces

$$\mathring{W}_p^{k,\varrho}(0,b) = \{ y \in W_p^{k,\varrho}(0,b) : y^{(i)}(0) = 0, \ i=0,\dots,k-1\}, \quad k \in \mathbb{N},$$

$$\mathring{C}^{k,\varrho}[0,b] = \{ y \in C^{k,\varrho}[0,b] : y^{(i)}(0) = 0, \ i=0,\dots,k\}, \quad k \in \mathbb{N}_0,$$

of $W_p^{k,\varrho}$ and $C^{k,\varrho}$, respectively. It is easy to check that

$$\|y\|_{\overset{\circ}{W}{}_p^{k,\varrho}} = \|x^{-\varrho}y^{(k)}\|_{L_p} \ , \quad \|y\|_{\overset{\circ}{C}{}^{k,\varrho}} = \|x^{-\varrho}y^{(k)}\|_{C^0} \tag{1.1.1}$$

are equivalent norms in $\overset{\circ}{W}{}_p^{k,\varrho}$ and $\overset{\circ}{C}{}^{k,\varrho}$, respectively. We set $L_p^{\varrho} = W_p^{0,\varrho} = \overset{\circ}{W}{}_p^{0,\varrho}$ ($\varrho \in \mathbb{R}$). Finally, let

$$\overset{\circ}{H}{}^{k,\lambda}[0,b] = \{ y \in H^{k,\lambda}[0,b] \ : \ y^{(i)}(0) = 0, \ i=0,\dots,k \} .$$

Then

$$\|y\|_{\overset{\circ}{H}{}^{k,\lambda}[0,b]} = \sup_{0 < x \leq t < b} |y^{(k)}(x)-y^{(k)}(t)|\,|x-t|^{-\lambda} \tag{1.1.2}$$

is an equivalent norm in the subspace $\overset{\circ}{H}{}^{k,\lambda}$ of $H^{k,\lambda}$.

<u>1.1.3.</u> For $l \in \mathbb{N}$, let

$$W_{p,l}^{k,\varrho}(0,b) = \{ y \in W_p^{k,\varrho}(0,b) \ : \ x^iD^iy \in W_p^{k,\varrho}(0,b), \ i=1,\dots,l \} ,$$

where the terms x^iD^iy are defined in the sense of distributions. $W_{p,l}^{k,\varrho}$ is a Banach space with the canonical norm

$$\|y\|_{W_{p,l}^{k,\varrho}(0,b)} = \sum_{0 \leq i \leq l} \|x^iD^iy\|_{W_p^{k,\varrho}(0,b)} .$$

Analogously

$$C_l^{k,\varrho}[0,b] = \{ y \in C^{k,\varrho}[0,b] \ : \ x^iD^iy \in C^{k,\varrho}[0,b], \ i=1,\dots,l \}$$

is a Banach space endowed with the canonical norm. Since x^iD^i is a linear combination of the terms $(xD)^j$ ($j \leq i$),

$$|y|_{W_{p,l}^{k,\varrho}} = \sum_{0 \leq i \leq l} \|(xD)^iy\|_{W_p^{k,\varrho}} \ , \quad |y|_{C_l^{k,\varrho}} = \sum_{0 \leq i \leq l} \|(xD)^iy\|_{C^{k,\varrho}} \tag{1.1.3}$$

are equivalent norms in $W_{p,l}^{k,\varrho}$ and $C_l^{k,\varrho}$, respectively. We shall also write $L_{p,l}^{\varrho}$ instead of $W_{p,l}^{0,\varrho}$.

<u>Lemma 1.1.1.</u> (i) For any $y \in C_l^{k,\varrho}[0,b]$,

$$D^j(x^iD^iy)|_{x=0} = j(j-1)\ \dots\ (j-i+1)(D^jy)(0) \ , \qquad j=0,\dots,k, \ i=1,\dots,l. \tag{1.1.4}$$

(ii) If $y \in W_{p,l}^{k,\varrho}(0,b)$, $k \in \mathbb{N}$, then (1.1.4) holds for $j=0,\dots,k-1$.

<u>Proof.</u> (i) Let $j=0$ and $i=1$. Since $xDy = Dxy-y$ on $(0,b]$ and

$$Dxy|_{x=0} = \lim_{x \to 0} xy/x = y(0) ,$$

we obtain $xDy|_{x=0} = 0$. Using the identities

$$x^iD^iy = (xD)(x^{i-1}D^{i-1}y) - (i-1)x^{i-1}D^{i-1}y, x\in(0,b] ,$$

we get $x^iD^iy|_{x=0} = 0$ for $i\geq 2$ by induction. Finally, by induction and the relations

$$y^{(j)}(0) = j! \lim_{x\to 0} [y(x) - \sum_{0\leq i<j} x^iy^{(i)}(0)/i!]\, x^{-j} ,$$

(1.1.4) holds for $j\geq 1$.

(ii) follows from (i) and the continuous embedding $W^{k,}_{p,l}\subset C^{k-1}_l$. □

Finally, we define the Banach space

$$H^{k,\lambda}_l[0,b] = \{y\in H^{k,\lambda}[0,b] : x^iD^iy\in H^{k,\lambda}[0,b] ,\ i=1,\dots,l\}$$

equipped with the canonical norm.

<u>1.1.4.</u> We introduce the closed subspaces

$$\mathring{W}^{k,\varrho}_{p,l}(0,b) = \{ y\in W^{k,\varrho}_{p,l}(0,b) : y^{(i)}(0)=0,\ i=0,\dots,k-1\} ,\ k\in \mathbb{N} ,$$

$$\mathring{C}^{k,\varrho}_l[0,b] = \{ y\in C^{k,\varrho}_l[0,b] : y^{(i)}(0)=0,\ i=0,\dots,k \} ,\ k\in \mathbb{N}_0 ,$$

of $W^{k,\varrho}_{p,l}$ and $C^{k,\varrho}_l$, respectively. It follows from 1.1.2 that

$$|y|_{\mathring{W}^{k,\varrho}_{p,l}} = \sum_{0\leq i\leq l} \|(xD)^iy\|_{\mathring{W}^{k,\varrho}_p}, |y|_{\mathring{C}^{k,\varrho}_l} = \sum_{0\leq i\leq l} \|(xD)^iy\|_{\mathring{C}^{k,\varrho}} \qquad (1.1.5)$$

are equivalent to the norms (1.1.3) in $\mathring{W}^{k,\varrho}_{p,l}$ and $\mathring{C}^{k,\varrho}_l$, respectively. Finally, let

$$\mathring{H}^{k,\lambda}_l[0,b] = \{y\in H^{k,\lambda}_l[0,b] : y^{(i)}(0)=0,\ i=0,\dots,k\} .$$

<u>1.2. The Euler operator</u>

<u>1.2.1.</u> With the operator A_o defined in (1.0.2), we associate the algebraic equation

$$\begin{aligned} P(\mu) = x^{-\mu}A_o(x^\mu) &= \mu(\mu-1)\dots(\mu-l+1) \\ &+ a_{l-1}(0)\mu\dots(\mu-l+2) + \dots + a_1(0)\mu + a_o(0) = 0 . \end{aligned} \qquad (1.2.1)$$

(1.2.1) is called the characteristic equation of A_o (and A) and its roots μ_i (i=1,...,l) are called the characteristic roots of A_o. (resp. A). Set $A_i = xD-\mu_i$. Then the following lemma is obvious.

Lemma 1.2.1. The Euler operator (1.0.2) has the representation

$$A_0 = A_l A_{l-1} \cdots A_1 . \tag{1.2.2}$$

Furthermore, the following commutation formulas hold:

$$A_0 x^{\varrho} = x^{\varrho}(A_l + \varrho) \cdots (A_1 + \varrho), \tag{1.2.3}$$

$$D^k A_0 = (A_l + k) \cdots (A_1 + k) D^k . \tag{1.2.4}$$

Let now μ_i $(i=1,\dots,\bar{l} \le l)$ be the different roots of equation (1.2.1) with multiplicities $r_i, r_1 + \dots + r_{\bar{l}} = l$.

Lemma 1.2.2. For $x \in (0,b]$, the homogeneous equation $A_0 y=0$ has a fundamental set of solutions of the form

$$y_{ij}(x) = x^{\mu_i}(\ln x)^j , \quad j=0,\dots,r_i-1, \quad i=1,\dots,\bar{l}. \tag{1.2.5}$$

Proof. By Lemma 1.2.1 and the equalities

$$(xD-\mu)^j [x^{\mu}(\ln x)^j] = j! x^{\mu}, \quad j \in \mathbb{N}, \ \mu \in \mathbb{C}, \tag{1.2.6}$$

we obtain $A_0 y_{ij}(x) = 0$ for $x>0$. Furthermore, the functions (1.2.5) are linearly independent on $(0,b]$. □

Let $\ker_p^{k,\varrho} A_0$ and $\ker^{k,\varrho} A_0$ be the kernel of the operator A_0 in $\mathring{W}^{k,\varrho}_{p,l}(0,b)$ and $\mathring{C}^{k,\varrho}_l[0,b]$, respectively. Using Lemma 1.2.2 we compute its dimension. Throughout this chapter let $\xi(p,k,\varrho)$ be the number of characteristic roots of A_0 satisfying $\operatorname{Re}\mu_i > -1/p + k + \varrho$ (with $1/p=0$ for $p=\infty$), $\xi^*(p,k,\varrho) = 0$ when $p<\infty$ and $\xi^*(\infty,k,\varrho)$ the number of different characteristic roots satisfying $\operatorname{Re}\mu_i = k+\varrho$, and

$$\xi^{**}(k,\varrho) = \begin{cases} 1 & \text{if} \quad \varrho>0 \text{ and } \mu_i=k+\varrho \text{ for some } i, \\ 0 & \text{otherwise}. \end{cases}$$

Lemma 1.2.3. $\dim \ker_p^{k,\varrho} A_0 = \xi(p,k,\varrho) + \xi^*(p,k,\varrho)$,

$\dim \ker^{k,\varrho} A_0 = \xi(\infty,k,\varrho) + \xi^{**}(k,\varrho)$.

Proof. For $p \in [1,\infty)$, we have $y_{ij} = x^{\mu_i}(\ln x)^j \in \ker_p^{k,\varrho} A_0$ if and only if $\operatorname{Re}\mu_i > -1/p + k + \varrho$. Furthermore,

$$y_{ij} \in \ker_{\infty}^{k,\varrho} A_0 \Longleftrightarrow \operatorname{Re}\mu_i > k+\varrho \quad \text{or } \operatorname{Re}\mu_i = k+\varrho, \ j=0,$$

$$y_{ij} \in \ker^{k,\varrho} A_0 \Longleftrightarrow \operatorname{Re}\mu_i > k+\varrho \quad \text{or} \quad \varrho>0, \ \mu_i = k+\varrho, \ j=0.$$

Let X (resp. Y) be one of the spaces $\overset{\circ}{W}{}^{k,\varrho}_{p,l}$ (resp. $\overset{\circ}{W}{}^{k,\varrho}_{p}$) and $\overset{\circ}{C}{}^{k,\varrho}_{l}$ (resp. $\overset{\circ}{C}{}^{k,\varrho}$). It remains to show that $y_o = \sum_M c_{ij} y_{ij} \in X$, $c_{ij} \in \mathbb{C}$, implies $c_{ij} = 0$ for any $(i,j) \in M$, where M denotes the set of pairs (i,j) satisfying $y_{ij} \bar{\in} X$. For fixed i, we set

$$B_{ij} = \prod_{1 \le m \le \bar{l}, m \ne i} (xD - \mu_m)^{r_m} (xD - \mu_i)^j, \quad j=0,\ldots,r_i-1 ,$$

and obtain

$$B_{ir_i-1} y_o = (r_i-1)! \prod_{m \ne i} (\mu_i - \mu_m)^{r_m} c_{ir_i-1} x^{\mu_i} \in Y$$

by (1.2.6) and the fact that the factors of B_{ij} commute. But the last relation implies $c_{ir_i-1}=0$. Applying successively the operators $B_{ij}(j=r_i-2,\ldots,0)$ to y_o, we get $c_{ij}=0$ for all j. □

<u>1.2.2.</u> For abbreviation, let $X^{k,\varrho}_p = \overset{\circ}{W}{}^{k,\varrho}_p$ (resp. $X^{k,\varrho}_{p,l} = \overset{\circ}{W}{}^{k,\varrho}_{p,l}$) and also $X^{k,\varrho}_\infty = \overset{\circ}{C}{}^{k,\varrho}$ (resp. $X^{k,\varrho}_{\infty,l} = \overset{\circ}{C}{}^{k,\varrho}_l$). In the sequel we drop the upper index $\cdot\varrho$ when $\varrho = 0$. By Lemma 1.1.1, we obviously have $A_o \in L(X^{k,\varrho}_{p,l}(0,b), X^{k,\varrho}_p(0,b))$. The following theorem is basic for the investigation of the Fuchsian operator.

<u>Theorem 1.2.4.</u> Under the assumption

$$\operatorname{Re}\mu_i \ne -1/p + k + \varrho \ , \quad i=1,\ldots,l \ , \tag{1.2.7}$$

$A_o \in L(X^{k,\varrho}_{p,l}(0,b), X^{k,\varrho}_p(0,b))$ is a right invertible operator[1] with kernel index $\xi(p,k,\varrho)$. Moreover, there exist right inverses $A^{-1}_{o,b}$ of A_o on $[0,b]$ and a constant $c > 0$ independent of b such that

$$\|A^{-1}_{o,b}\|_{L(X^{k,\varrho}_p(0,b), X^{k,\varrho}_{p,l}(0,b))} \le c \ , \quad b > 0 \ . \tag{1.2.8}$$

We first prove Theorem 1.2.4 for k=0 and a first order operator $B = xD - \mu$, $\mu \in \mathbb{C}$. Then condition (1.2.7) reads

$$\operatorname{Re}\mu \ne -1/p + \varrho \tag{1.2.9}$$

and we have the cases

(i) $\operatorname{Re}\mu < -1/p + \varrho$, (ii) $\operatorname{Re}\mu > -1/p + \varrho$.

---

1) $A \in L(X,Y)$ is called right invertible if there exists $B \in L(Y,X)$ such that $AB=I$ (= identity operator in Y). B is called a right inverse of A.

Lemma 1.2.5. Under the assumption (1.2.9), the equation

$$By = f, \ f \in X_p^{0,\varrho}(0,b), \tag{1.2.10}$$

has always a solution $y \in X_{p,1}^{0,\varrho}(0,b)$ satisfying the estimate

$$|\mathrm{Re}\,\mu + 1/p - \varrho| \ \|y\|_{X_p^{0,\varrho}(0,b)} \le \|f\|_{X_p^{0,\varrho}(0,b)} . \tag{1.2.11}$$

Proof. 1. Let $\varrho = 0$. In case (i) we set

$$y(x) = B_b^{-1}(x) = x^{\mu}\int_0^x f(z)z^{-\mu-1}dz, \qquad x > 0 . \tag{1.2.12}$$

By differentiation it is easily seen that (1.2.12) is a solution of equation (1.2.10) on $(0,b]$. Now we verify that $y \in X_p^0(0,b)$.

Let $p < \infty$. Setting $\varphi(x,z) = |f(xz)|z^{-r-1}$, $r = \mathrm{Re}\,\mu$, and applying an inequality due to Minkowski (cf. Stein [1, Appendix A.1]), we obtain

$$\left[\int_0^b |y(x)|^p dx\right]^{1/p} = \left[\int_0^b \left|x^{\mu}\int_0^x f(z)z^{-\mu-1}dz\right|^p dx\right]^{1/p}$$

$$\le \left[\int_0^b \left(\int_0^1 \varphi(x,z)dz\right)^p dx\right]^{1/p} \le \int_0^1 \left(\int_0^b \varphi(x,z)^p dx\right)^{1/p} dz$$

$$= \int_0^1 z^{-r-1}\left(\int_0^b |f(x\,z)|^p dx\right)^{1/p} dz \le \|f\|_{L_p(0,b)} \int_0^1 z^{-r-1-1/p}dz$$

$$= |r + 1/p|^{-1} \|f\|_{L_p(0,b)} . \tag{1.2.13}$$

For $X_\infty^0 = L_\infty$, we have

$$\operatorname*{ess\,sup}_{0 \le x \le b} |y(x)| \le \operatorname*{ess\,sup}_{0 \le x \le b} x^r \int_0^x \|f\|_{L_\infty(0,b)} z^{-r-1}dz$$

$$\le |r|^{-1} \|f\|_{L_\infty(0,b)} . \tag{1.2.14}$$

For $X_\infty^0 = \mathring{C}^0$, we deduce that the function y defined by (1.2.12) for $x > 0$ and $y(0)=0$ is continuous on $[0,b]$. Indeed, since $f(0)=0$, l'Hospital's rule yields

$$\lim_{x \to 0+} \left(\int_0^x |f(z)|\, z^{-r-1}dz\right)/x^{-r} = 0$$

which implies $y(x) \to 0$ as $x \to 0+$. We add a remark which is needed in the proof for $\varrho > 0$: The function y defined by (1.2.12) and $y(0) = - f(0)/\mu$ is a continuous solution of the equation $By = f \in C^0[0,b]$ satisfying inequality (1.2.11), since $B_b^{-1}\{f(x) - f(0)\} \in \mathring{C}^0[0,b]$.

By (1.2.13) and (1.2.14), we now obtain estimate (1.2.11). Moreover, since y, $By \in X_p^o(0,b)$, it follows that $y \in X_{p,1}^o(0,b)$.

2. Let $\rho = 0$ again. In case (ii) we set

$$y(x) = B_b^{-1} f(x) = x^{\mu} \int_b^x f(z) z^{-\mu-1} dz, \quad x > 0 , \tag{1.2.15}$$

which is a solution of (1.2.1o) on $(0,b]$. To verify $y \in X_p^o(0,b)$ for $p < \infty$, we use the function φ and Minkowski's inequality as above and obtain

$$\|y\|_{L_p(0,b)} = \left[\int_0^b |x^{\mu} \int_b^x f(z) z^{-\mu-1} dz|^p dx\right]^{1/p}$$

$$\le \left[\int_0^b \left(\int_1^{b/x} \varphi(x,z) dz\right)^p dx\right]^{1/p} \le \int_1^{\infty} \left(\int_0^b \varphi(x,z)^p dx\right)^{1/p} dz$$

$$= \int_1^{\infty} z^{-r-1} \left(\int_0^b |f(xz)|^p dx\right)^{1/p} \le |r + 1/p|^{-1} \|f\|_{L_p(0,b)} .$$

For $X_{\infty}^o = L_{\infty}$, we get (1.2.14) again. As in 1., one can handle the case $X_{\infty}^o = \mathring{C}^o$ and complete the proof of the lemma. Note that (1.2.15) also yields a continuous solution of $By = f \in C^o[0,b]$ which satisfies estimate (1.2.11).

3. For $\rho \neq 0$, we pass to the operator $x^{-\rho} B x^{\rho} = xD - \mu + \rho$. Using the arguments from 1. and 2. for this operator and the relations $L_p^{\rho} = x^{\rho} L_p$, $\mathring{C}^{o,\rho} = x^{\rho} C^o$, we conclude the lemma. □

<u>Corollary 1.2.6.</u> Under hypothesis (1.2.9), the right inverses B_b^{-1} defined in (1.2.12) and (1.2.15) satisfy the estimate

$$\|B_b^{-1} f\|_{X_{p,1}^{o,\rho}(0,b)} \le c_{\mu} \|f\|_{X_p^{o,\rho}(0,b)}, \quad f \in X_p^{o,\rho}(0,b) , \tag{1.2.16}$$

where $c_{\mu} = |\mathrm{Re}\,\mu + 1/p - \rho|^{-1}(1 + |\mu| + |\mathrm{Re}\,\mu + 1/p - \rho|)$.

<u>Proof.</u> From (1.2.11) and the obvious estimate

$$\|By\|_{X_p^{o,\rho}} + |\mu| \, \|y\|_{X_p^{o,\rho}} \ge \|xDy\|_{X_p^{o,\rho}}$$

it follows that

$$(|\mathrm{Re}\,\mu + 1/p - \rho| + |\mu|) \, \|By\|_{X_p^{o,\rho}} \ge |\mathrm{Re}\,\mu + 1/p - \rho| \, \|xDy\|_{X_p^{o,\rho}} .$$

Together with (1.2.11) this implies (1.2.16). □

1.2.3. Proof of Theorem 1.2.4 for k=0. We consider the equation

$$A_0 y = f \ , \quad f \in X_p^{0,\varrho}(0,b) \ , \tag{1.2.17}$$

where A_0 has the representation (1.2.2). Suppose (1.2.7) holds with k=0. Let $A_{i,b}^{-1}$ be the right inverses of the first order operators A_i constructed in (1.2.12) or (1.2.15) and set $A_{0,b}^{-1} = A_{1,b}^{-1} \dots A_{l,b}^{-1}$. By Lemma 1.2.5, (1.2.17) has the solution $y = A_{0,b}^{-1}f$ in $X_p^{0,\varrho}(0,b)$ and, since $A_i \dots A_1 y \in X_p^{0,\varrho} (i=1,\dots,l)$, $y \in X_{p,l}^{0,\varrho}(0,b)$. We show by induction on l that the estimate

$$|y|_{X_{p,l}^{0,\varrho}(0,b)} = |A_{0,b}^{-1}f|_{X_{p,l}^{0,\varrho}(0,b)} \le c \, \|f\|_{X_p^{0,\varrho}(0,b)} \ , \qquad c = c_{\mu_1} \dots c_{\mu_l} \ , \tag{1.2.18}$$

holds, where $|\cdot|$ is the norm (1.1.5) and c_μ the constant defined in Corollary 1.2.6.

By Corollary 1.2.6, (1.2.18) holds for l=1. Now suppose the assertion is true for l > 1 and consider an operator BA_0, where $B = xD - \mu$ and μ satisfies (1.2.9). Setting $y = A_{0,b}^{-1}B_b^{-1}f$, by assumption we then obtain

$$|y|_{X_{p,l}^{0,\varrho}(0,b)} \le \sum_{0 \le i \le l} \left\{ \|(xD)^{i+1}y\|_{X_p^{0,\varrho}(0,b)} + \|(xD)^i y\|_{X_p^{0,\varrho}(0,b)} \right\}$$

$$\le c_\mu \sum_{0 \le i \le l} \|B(xD)^i y)\|_{X_p^{0,\varrho}(0,b)} \le c c_\mu \|A_0 By\|_{X_p^{0,\varrho}(0,b)}$$

which proves the assertion for Euler operators of order l+1. Thus inequality (1.2.8) is proved. The formula for the kernel index of A_0 is an immediate consequence of Lemma 1.2.3. □

Remark 1.2.7. Exactly as in 1.1.1 and 1.1.3 one can define the spaces $L_p^\varrho(0,\infty)$ and $L_{p,l}^\varrho(0,\infty)$. If (1.2.7) holds with k=0, then the Euler operator $A_0 \in L(L_{p,l}^\varrho(0,\infty), L_p^\varrho(0,\infty))$ is invertible and its inverse A_0^{-1} satisfies

$$\|A_0^{-1}\|_{L(L_p^\varrho(0,\infty), L_{p,l}^\varrho(0,\infty))} \le c = c_{\mu_1} \dots c_{\mu_l} \ . \tag{1.2.19}$$

Indeed, an inspection of the corresponding proofs shows that Lemma 1.2.5 and Corollary 1.2.6 remain valid for $X_p^{0,\varrho} = L_p^\varrho$ and $b = \infty$. Thus we obtain as above that A_0 has a right inverse A_0^{-1} satisfying (1.2.19), and the kernel of A_o is trivial by Lemma 1.2.2.

<u>1.2.4. Proof of Theorem 1.2.4 for $k>0$.</u> By (1.2.4) we have the commutative diagram

$$\begin{array}{ccccc} A_o : & X^{k,\varrho}_{p,l}(0,b) & \longrightarrow & X^{k,\varrho}_{p}(0,b) & \\ & \downarrow D^k & & \downarrow D^k & \qquad (1.2.2o) \\ A_k : & X^{o,\varrho}_{p,l}(0,b) & \longrightarrow & X^{o,\varrho}_{p}(0,b) & , \end{array}$$

where A_k is an Euler operator with the characteristic roots μ_i-k ($i=1,\dots,l$). Using the norms (1.1.1) and (1.1.5), we observe that D^k are isomorphisms in (1.2.2o) with norms independent of b. On the other hand, it follows from 1.2.3 that A_k is right invertible with kernel index $\xi(p,k,\varrho)$ and has a right inverse with norm independent of b. Hence, by (1.2.2o), Theorem 1.2.4 holds for $k\in\mathbb{N}$. □

<u>1.3. The Fuchsian operator (I)</u>

<u>1.3.1.</u> We first extend Theorem 1.2.4 to the Fuchsian operator A defined in (1.o.1). For its coefficients a_i, assume that

$$a_i \in C^k[0,b]\ ,\ i=0,\dots,l-1\ . \qquad (1.3.1)$$

Retaining the notation of 1.2.2, we then have $A\in L(X^{k,\varrho}_{p,l}(0,b),\ X^{k,\varrho}_{p}(0,b))$. This is obvious when $k=0$ or $\varrho=0$ and a consequence of the continuous embeddings $X^{j,\varrho}_p\subset X^{j-1,\varrho}_p$ ($j\in\mathbb{N}$) when $k\in\mathbb{N}$ and $\varrho>0$. But the latter relation follows from the inequality

$$|x^{-\varrho}y^{(j-1)}(x)| = |x^{-\varrho}\int_0^x y^{(j)}(z)dz| \le \int_0^x |z^{-\varrho}y^{(j)}(z)|dz,\ 0<x\le b\ ,$$

if $y\in X^{j,\varrho}_p(0,b)$.

<u>Theorem 1.3.1.</u> Assume (1.3.1).

(i) Under hypothesis (1.2.7), $A\in L(X^{k,\varrho}_{p,l}(0,b),X^{k,\varrho}_{p}(0,b))$ is a right invertible operator with kernel index $\xi(p,k,\varrho)$.

(ii) If (1.2.7) is violated, then $A\in L(X^{k,\varrho}_{p,l}(0,b),X^{k,\varrho}_{p}(0,b))$ is not normally solvable.

<u>Proof.</u> (i) Setting $A' = A - A_o$, we obtain

$$\|A'\|_{L(X^{k,\varrho}_{p,l}(0,\varepsilon),X^{k,\varrho}_{p}(0,\varepsilon))} \to 0 \quad \text{as} \quad \varepsilon\to 0, \qquad (1.3.2)$$

since the coefficients of the terms $x^i D^i$ in A' vanish at the origin and

$$\| D^j \|_{L(X_p^{k,\varrho}(0,\varepsilon), X_p^{0,\varrho}(0,\varepsilon))} \to 0 \quad (j=0,\dots,k-1) \quad \text{as } \varepsilon \to 0 .$$

By Theorem 1.2.4, $A_o \in L(X_{p,1}^{k,\varrho}(0,\varepsilon), X_p^{k,\varrho}(0,\varepsilon))$ is right invertible with kernel index $\xi(p,k,\varrho)$ and has a right inverse $A_{o,\varepsilon}^{-1}$ with norm independent of ε. Furthermore, by (1.3.2) there exists a sufficiently small number $\varepsilon > 0$ such that $T = A'A_{o,\varepsilon}^{-1} + I$ is invertible in $X_p^{k,\varrho}(0,\varepsilon)$. Then $A_{o,\varepsilon}^{-1} T^{-1}$ is a right inverse of A. Since

$$\dim \ker A = \operatorname{ind} A = - \operatorname{ind} A_{o,\varepsilon}^{-1} = \operatorname{ind} A_o = \xi(p,k,\varrho) ,$$

assertion (i) is valid for $b = \varepsilon$. Finally, by the classical existence and uniqueness theorem for linear ordinary differential equations (cf. Coddington and Levinson [1, Chap. 3]), each solution $y \in X_{p,1}^{k,\varrho}(0,\varepsilon)$ of the equation $Ay = f \in X_p^{k,\varrho}(0,b)$ can uniquely be extended to a solution in $X_{p,1}^{k,\varrho}(0,b)$. Thus the proof of (i) is complete.

(ii) Let $\operatorname{Re} \mu_i = -1/p+k+\varrho$ for some i. We argue by contradiction and assume that A is normally solvable. Since obviously $\dim \ker A \leq 1$, A is then a Φ_+-operator. Consider the operators $A(\varepsilon) = A_o(\varepsilon) + A - A_o$, $\varepsilon \in \mathbb{R}$, where

$$A_o(\varepsilon) = (xD-\mu_1-\varepsilon) \dots (xD-\mu_1-\varepsilon) .$$

We observe that

$$\| A(\varepsilon) - A \|_{L(X_{p,1}^{k,\varrho}(0,b), X_p^{k,\varrho}(0,b))} \to 0 \quad \text{as } \varepsilon \to 0 ,$$

so that $A(\varepsilon)$ is a small perturbation of $A = A(0)$ with respect to the operator norm. Moreover, there exists a sufficiently small number $c > 0$ such that the operators $A(\varepsilon)$ satisfy condition (1.2.7) for all $\varepsilon \in [-c,c] \setminus \{0\}$. Consequently, by (i), $A(\varepsilon) \in L(X_{p,1}^{k,\varrho}(0,b), X_p^{k,\varrho}(0,b))$ are Fredholm operators and

$$\operatorname{ind} A(\varepsilon) = \text{const} < \operatorname{ind} A(\bar{\varepsilon}) = \text{const}, \quad \varepsilon \in [-c,0), \quad \bar{\varepsilon} \in (0,c] .$$

This contradicts the stability of the index of a Φ_+-operator with respect to small perturbations (cf. Goldberg [1, Th. V.1.6]). □

1.3.2. In order to avoid the exceptional values for the characteristic roots μ_i in Theorem 1.3.1, we now consider the Fuchsian operator

(1.o.1) as a map of $X^{k,\varrho}_{p,l}$ into $X^{k,\bar\varrho}_{p}$ with $\bar\varrho > \varrho$ and the domain of definition

$$\{y \in X^{k,\varrho}_{p,l}(0,b) : Ay \in X^{k,\bar\varrho}_{p}(0,b)\} . \qquad (1.3.3)$$

Under assumption (1.3.1), this is a closed operator which is always surjective. Indeed, if $\bar\varrho$ is sufficiently close to ϱ, then the equation $Ay=f$ has a solution $y \in X^{k,\bar\varrho}_{p,l} \subset X^{k,\varrho}_{p,l}$ for every right-hand side $f \in X^{k,\bar\varrho}_{p}$ in view of Theorem 1.3.1. Under a somewhat stronger assumption on the smoothness of the coefficients of A, it is possible to compute the kernel index.

<u>Theorem 1.3.2.</u> Assume $\bar\varrho > \varrho$ and

$$a_i \in C^{k,\sigma}[0,b] , \ i=0,\dots,l-1, \text{ for some } \sigma > 0 . \qquad (1.3.4)$$

Then the operator $A : X^{k,\varrho}_{p,l}(0,b) \to X^{k,\bar\varrho}_{p}(0,b)$ with domain of definition (1.3.3) is right invertible and has the kernel index

$$\dim \ker A = \begin{cases} \xi(p,k,\varrho) + \xi^{*}(p,k,\varrho) & \text{if } X^{k,\varrho}_{p} = \mathring{W}^{k,\varrho}_{p} , \\ \xi(\infty,k,\varrho) + \xi^{**}(k,\varrho) & \text{if } X^{k,\varrho}_{\infty} = \mathring{C}^{k,\varrho} . \end{cases}$$

<u>Proof.</u> It is sufficient to consider the case when $\bar\varrho - \varrho$ is small. Thus we can assume that $0 < \bar\varrho - \varrho < \sigma$ and $\mathrm{Re}\,\mu_i \neq -1/p+k+\bar\varrho$ $(i=1,\dots,l)$. Then estimate (1.2.8) holds with ϱ replaced by $\bar\varrho$. Consequently,

$$\| A^{-1}_{0,b} \|_{L(X^{k,\bar\varrho}_{p}(0,b),X^{k,\varrho}_{p,l}(0,b))} \le c, \ b > 0 .$$

On the other hand, (1.3.4) implies that

$$\| A-A_0 \|_{L(X^{k,\varrho}_{p,l}(0,b),X^{k,\bar\varrho}_{p}(0,b))} \to 0 \quad \text{as } b \to 0 .$$

Applying these two estimates and Lemma 1.2.3, we obtain the result by the same arguments as in the proof of Theorem 1.3.1(i). □

<u>Remark 1.3.3.</u> The following example shows that the above formula for the kernel index does not hold, in general, when the coefficients of A are only continuous and k=0. Consider the operator $A = xD-g(x)$ on $[0,1/2]$, where $g(x) = 1/\ln x$ for $x > 0$ and $g(0) = 0$. The corresponding Euler operator $A_0 = xD$ has the characteristic root 0. However, the equation $Ay = 0$ has the general solution $y = c \ln x$, $c \in \mathbb{C}$, on $(0,1/2]$,

and $\ln x \in L_\infty(0,1/2)$. Therefore the kernel of the surjective operator $A : L_\infty(0,1/2) \to L_\infty^{\bar\rho}(0,1/2)$, $\bar\rho > 0$, is trivial so that Theorem 1.3.2 is not valid in this case.

1.4. The Fuchsian operator (II)

1.4.1. In order to study the Fuchsian operator A defined by (1.o.1) in the spaces $W_p^{k,\rho}$ and $C^{k,\rho}$, we need some simple algebraic considerations. Let $k \in \mathbb{N}_0$ and $y \in W_{p,1}^{k+1,\rho}(0,b)$ (resp. $\in C_1^{k,\rho}[0,b]$) be a solution of the equation $Ay=f$. By Lemma 1.1.1, with this equation we can associate the system of linear equation

$$(Ay)^{(j)}(0) = f^{(j)}(0)\ , \quad j=0,\dots,k\ , \tag{1.4.1}$$

in the unknowns $y(0),\dots,y^{(k)}(0)$. (1.4.1) can be written

$$\mathcal{A}_k Y_k = F_k \tag{1.4.2}$$

with $Y_k = (y_j)_0^k$, $y_j = y^{(j)}(0)$, $F_k = (f^{(j)}(0))_0^k$ and some matrix operator $\mathcal{A}_k$ in $\mathbb{C}^{k+1}$ the kernel of which is denoted by n(k). System (1.4.2) is solvable if and only if F_k satisfies the solvability condition

$$F_k \perp \ker \mathcal{A}_k^* \ , \tag{1.4.3}$$

i.e. F_k is orthogonal (in $\mathbb{C}^{k+1}$) to the kernel of the adjoint of $\mathcal{A}_k$. We observe that $\mathcal{A}_k$ is in lower triangular form and the diagonal element in the (k+1)-th row is given by the zero order term in the Taylor expansion of $x^{-k}A(x^k)$ at $x=0$, which is P(k) with P defined in (1.2.1). Let $k_0 = \min\{k \in \mathbb{N}_0 : P(j) \neq 0,\ j=k,k+1,\dots\}$. Then we obviously have

Lemma 1.4.1. (i) $\mathcal{A}_k$ is invertible if and only if $P(j) \neq 0$, $j=0,\dots,k$.
(ii) For $k \geq k_0$, condition (1.4.3) is equivalent to

$$F_{k_0-1} \perp \ker \mathcal{A}_{k_0-1}^* \ . \tag{1.4.4}$$

Furthermore, $n(k) = n(k_0-1)$ for all $k \geq k_0$, where (1.4.4) is void and $n(k_0-1)$ is defined to be 0 if $k_0=0$.

1.4.2. Suppose the coefficients of A satisfy (1.3.1). Furthermore, let

$$\rho \in [0,1/p) \text{ (resp. } \rho=0) \text{ when } k \in \mathbb{N} \text{ and } p<\infty \text{ (resp. } p=\infty). \tag{1.4.5}$$

Then $A \in L(W_{p,1}^{k,\rho}(0,b),\ W_p^{k,\rho}(0,b))$ in view of the continuous embeddings $W_p^j \subset W_p^{j-1,\rho}$, $j \in \mathbb{N}$.

Theorem 1.4.2. Assume (1.3.1) and (1.4.5).

(i) Under hypothesis (1.2.7), $A \in L(W^{k,\varrho}_{p,1}(0,b), W^{k,\varrho}_{p}(0,b))$ is a Fredholm operator with kernel index $\zeta(p,k,\varrho) + n(k-1)$ and deficiency index $n(k-1)$. Moreover, for $f \in W^{k,\varrho}_{p}(0,b)$, the equation $Ay = f$ has a solution in $W^{k,\varrho}_{p,1}(0,b)$ if and only if

$$F_{k-1} \perp \ker \mathcal{A}^{*}_{k-1} \,. \tag{1.4.6}$$

(ii) If condition (1.2.7) is violated, then $A \in L(W^{k,\varrho}_{p,1}(0,b), W^{k,\varrho}_{p}(0,b))$ is not normally solvable.

Proof. Note that the case $k=0$ is already contained in Theorem 1.3.1. Let $k \in \mathbb{N}$.

(i) By Theorem 1.3.1(i), $A \in L(\mathring{W}^{k,\varrho}_{p,1}(0,b), \mathring{W}^{k,\varrho}_{p}(0,b))$ is a surjective Fredholm operator with index $\zeta(p,k,\varrho)$. Let $\mathcal{P}_k$ be the $(k+1)$-dimensional linear space of polynomials of degree $\leq k$. Since

$$\mathring{W}^{k,\varrho}_{p} \dotplus \mathcal{P}_{k-1} = W^{k,\varrho}_{p} \quad , \quad \mathring{W}^{k,\varrho}_{p,1} \dotplus \mathcal{P}_{k-1} = W^{k,\varrho}_{p,1} \quad ,$$

$A \in L(W^{k,\varrho}_{p,1}(0,b), W^{k,\varrho}_{p}(0,b))$ is also a Fredholm operator with index $\zeta(p,k,\varrho)$ (cf. Goldberg [1, Lemma V.1.5]).

Next, we observe that $\dim W^{k,\varrho}_{p}/A(W^{k,\varrho}_{p,1}) \geq n(k-1)$, since the solvability of the equation $Ay = f \in W^{k,\varrho}_{p}$ in $W^{k,\varrho}_{p,1}$ implies (1.4.6). Conversely, suppose (1.4.6) is satisfied. Then the linear system $\mathcal{A}_{k-1}Y_{k-1} = F_{k-1}$ has a solution $Y_{k-1} = (y_j)_{0}^{k-1}$. Consider the polynomial $\beta(x) = y_0 + \dots + y_{k-1}x^{k-1}/(k-1)!$. Since $A\beta - f \in \mathring{W}^{k,\varrho}_{p}$, Theorem 1.3.1(i) implies that there exists $\bar{y} \in \mathring{W}^{k,\varrho}_{p,1}$ such that $A\bar{y} = f - A\beta$. Hence, $y = \bar{y} + \beta \in W^{k,\varrho}_{p,1}$ is a solution of $Ay = f$. Thus $\dim W^{k,\varrho}_{p}/A(W^{k,\varrho}_{p,1}) = n(k-1)$ and (i) is proved.

(ii) Assume that (1.2.7) is violated and A is normally solvable. By Lemma V.1.5 in Goldberg [1], then the restriction $A|_{\mathring{W}^{k,\varrho}_{p,1}}$ would be normally solvable in $\mathring{W}^{k,\varrho}_{p}$ which contradicts Theorem 1.3.1(ii). □

As an immediate consequence of Lemma 1.4.1 and Theorem 1.4.2, we obtain

Corollary 1.4.3. Assume (1.3.1), $k \in \mathbb{N}$ and (1.4.5).

(i) $A \in L(W^{k,\varrho}_{p,1}(0,b), W^{k,\varrho}_{p}(0,b))$ is invertible if and only if

$$\mu_1 \neq 0,1,\dots,k-1 \tag{1.4.7}$$

and

$$\operatorname{Re}\mu_i < -1/p + k + \varrho, \quad i=1,\dots,l \, . \tag{1.4.8}$$

This operator is surjective if and only if (1.4.7) holds.

(ii) For each solution Y_{k-1} of the system $\mathcal{A}_{k-1}Y_{k-1} = F_{k-1}$, there exists exactly one solution $y \in W^{k,\varrho}_{p,l}$ of the equation $Ay = f \in W^{k,\varrho}_{p}$ satisfying $y^{(j)}(0) = y_j$ ($j=0,\dots,k-1$) if and only if (1.4.8) holds.

1.4.3. Under the assumptions (1.3.1) and

$$0 \le \varrho < 1 \quad \text{when} \quad k > 0 \, , \tag{1.4.9}$$

the Fuchsian operator A is a continuous map of $C^{k,\varrho}_l$ into $C^{k,\varrho}$. Using Theorem 1.3.1 and the relations $\overset{\circ}{C}{}^{k,\varrho} \dotplus \mathcal{P}_k = C^{k,\varrho}$, $\overset{\circ}{C}{}^{k,\varrho}_l \dotplus \mathcal{P}_k = C^{k,\varrho}_l$, by the same arguments as above we obtain

Theorem 1.4.4. Assume (1.3.1) and (1.4.9).

(i) Under the condition

$$\operatorname{Re}\mu_i \neq k+\varrho, \quad i=1,\dots,l \, , \tag{1.4.10}$$

$A \in L(C^{k,\varrho}_l[0,b], C^{k,\varrho}[0,b])$ is a Fredholm operator with nullity $\xi(\infty,k,\varrho) + n(k)$ and deficiency $n(k)$. Moreover, for $f \in C^{k,\varrho}[0,b]$, the equation $Ay = f$ is solvable in $C^{k,\varrho}_l[0,b]$ if and only if (1.4.3) holds.

(ii) If condition (1.4.10) is violated, then $A \in L(C^{k,\varrho}_l[0,b], C^{k,\varrho}[0,b])$ is not normally solvable.

Remark 1.4.5. Suppose condition (1.4.10) is satisfied for $\varrho = 0$. Then it follows from the considerations in 1.4.1 that (1.4.3) is equivalent to (1.4.6), since $P(k) \neq 0$. In particular, A is then always surjective in C^0. Furthermore, A is surjective in C^k if and only if (1.4.7) holds.

1.4.4. For $\bar\varrho > \varrho$, we consider the closed operators $A : W^{k,\varrho}_{p,l} \to W^{k,\bar\varrho}_{p}$ and $A : C^{k,\varrho}_l \to C^{k,\bar\varrho}$ with domains of definition

$$D^{\varrho,\bar\varrho}_{p,k}(A) = \{ y \in W^{k,\varrho}_{p,l}(0,b) : Ay \in W^{k,\bar\varrho}_{p}(0,b) \} \, , \tag{1.4.11}$$

$$D^{\varrho,\bar\varrho}_{k}(A) = \{ y \in C^{k,\varrho}_l[0,b] : Ay \in C^{k,\bar\varrho}[0,b] \} \, . \tag{1.4.12}$$

Using Theorem 1.3.2 and the same arguments as in 1.4.2 and 1.4.3, one can prove

Theorem 1.4.6. Under the hypotheses (1.3.4), (1.4.5) and $\bar\varrho > \varrho$, the operator $A : W^{k,\varrho}_{p,1}(0,b) \to W^{k,\bar\varrho}_p(0,b)$ with domain of definition (1.4.11) is Fredholm with nullity $\xi(p,k,\varrho) + \xi^*(p,k,\varrho) + n(k-1)$ and deficiency $n(k-1)$. Moreover, for $f \in W^{k,\bar\varrho}_p(0,b)$, the equation $Ay=f$ is solvable in $W^{k,\varrho}_{p,1}(0,b)$ if and only if (1.4.6) holds.

Theorem 1.4.7. Under the hypotheses (1.3.4), (1.4.9) and $\bar\varrho > \varrho$, the operator $A : C^{k,\varrho}_1[0,b] \to C^{k,\bar\varrho}[0,b]$ with domain of definition (1.4.12) is Fredholm with nullity $\xi(\infty,k,\varrho) + \xi^{**}(k,\varrho) + n(k)$ and deficiency $n(k)$. Furthermore, for $f \in C^{k,\bar\varrho}[0,b]$, the equation $Ay = f$ has a solution in $C^{k,\varrho}_1[0,b]$ if and only if (1.4.3) is satisfied.

1.5. Smoothness and Hölder continuity of solutions

1.5.1. We first state the following regularity result.

Theorem 1.5.1. Assume (1.3.1).

(i) Under the hypotheses (1.4.5) and $\operatorname{Re}\mu_i \,\bar\in\, [-1/p, -1/p+k+\varrho]$, $i=1,\dots,l$, $y \in W^{0}_{p,1}(0,b)$ and $Ay \in W^{k,\varrho}_p(0,b)$ imply $y \in W^{k,\varrho}_{p,1}(0,b)$.

(ii) Under the hypotheses (1.4.9) and $\operatorname{Re}\mu_i \,\bar\in\, [0,k+\varrho]$, $i=1,\dots,l$, $y \in C^0_1[0,b]$ and $Ay \in C^{k,\varrho}[0,b]$ imply $y \in C^{k,\varrho}_1[0,b]$.

Proof. Let us show (i). The proof of (ii) is analogous. We obviously have $\{y \in W^{k,\varrho}_{p,1} : Ay = 0\} \subset \{y \in W^0_{p,1} : Ay = 0\}$ and $A(W^{k,\varrho}_{p,1}) \subset A(W^0_{p,1}) \cap W^{k,\varrho}_p$. On the other hand, Theorem 1.4.2 implies $A \in L(W^0_{p,1}, W^0_p)$ and $A \in L(W^{k,\varrho}_{p,1}, W^{k,\varrho}_p)$ are both Fredholm operators with index $\xi(p,k,\varrho)$, since by assumption $\xi(p,0,0) = \xi(p,k,\varrho)$. Furthermore, since $W^{k,\varrho}_p$ is dense in W^0_p, the topological complement of $A(W^0_{p,1})$ in W^0_p can be chosen as a subspace of $W^{k,\varrho}_p$ and

$$\dim W^0_p/A(W^0_{p,1}) = \dim W^{k,\varrho}_p/A(W^0_{p,1}) \cap W^{k,\varrho}_p$$

(cf. Przeworska-Rolewicz and Rolewicz [1, Chap. B.III, Lemma 2.3]). Thus the kernels of A in $W^0_{p,1}$ and $W^{k,\varrho}_{p,1}$ as well as the linear spaces $A(W^0_{p,1}) \cap W^{k,\varrho}_p$ and $A(W^{k,\varrho}_{p,1})$ must coincide. □

1.5.2. We now consider the operator (1.0.1) in the space $C^\infty[0,b]$. Let k_0 as in 1.4.1 and

$$a_i \in C^\infty[0,b] \quad , \quad i=0,\dots,l-1 . \tag{1.5.1}$$

<u>Theorem 1.5.2.</u> Under hypothesis (1.5.1), $A \in L(C^{\infty}[0,b])$ is a Fredholm operator with index 0 and nullity $n(k_0-1)$. Moreover, for $f \in C^{\infty}[0,b]$, the equation $Ay = f$ is solvable in $C^{\infty}[0,b]$ if and only if (1.4.4) holds. In particular, A is invertible in $C^{\infty}[0,b]$ if and only if $\mu_i \bar{\in} N_0$ for $i=1,\dots,l$.

<u>Proof.</u> Let $\bar{k} \in N$ such that $\mathrm{Re}\,\mu_i < \bar{k}$ $(i=1,\dots,l)$. By Theorem 1.4.4 for $k = \bar{k}$ and $\varrho = 0$ and Lemma 1.4.1, $A \in L(C_l^{\bar{k}}[0,b], C^{\bar{k}}[0,b])$ is a Fredholm operator with nullity $n(k_0-1)$ and solvability conditions (1.4.4). Taking $\bar{k} \to \infty$, we obtain the result. □

As in the proof of Theorem 1.5.1, from Theorems 1.4.4 and 1.5.2 one can deduce

<u>Corollary 1.5.3.</u> Under the hypotheses (1.5.1) and $\mathrm{Re}\,\mu_i < 0$, $i=1,\dots,l$, $y \in C_l^0[0,b]$ and $Ay \in C^{\infty}[0,b]$ imply $y \in C^{\infty}[0,b]$.

<u>1.5.3.</u> Using Theorem 1.3.1 and a regularity argument, we finally study the Fuchsian operator in Hölder spaces.

<u>Theorem 1.5.4.</u> Assume $a_i \in H^{k,\lambda}[0,b]$, $i=0,\dots,l-1$.

(i) Under the hypothesis

$$\mathrm{Re}\,\mu_i \neq k+\lambda, \quad i=1,\dots,l, \qquad (1.5.2)$$

$A \in L(H_l^{k,\lambda}[0,b], H^{k,\lambda}[0,b])$ is a Fredholm operator with nullity $\xi(\infty,k,\lambda) + n(k)$ and deficiency $n(k)$. Moreover, for $f \in H^{k,\lambda}[0,b]$, the equation $Ay = f$ is solvable in $H_l^{k,\lambda}[0,b]$ if and only if (1.4.3) holds.

(ii) If condition (1.5.2) is violated, then $A \in L(H_l^{k,\lambda}[0,b], H^{k,\lambda}[0,b])$ is not normally solvable.

<u>Proof.</u> (i) From the proof of Theorem 1.4.2(i) and the relations $\mathring{H}^{k,\lambda} \dotplus \mathcal{P}_k = H^{k,\lambda}$, $\mathring{H}_l^{k,\lambda} \dotplus \mathcal{P}_k = H_l^{k,\lambda}$ we see that it suffices to show that $A \in L(\mathring{H}_l^{k,\lambda}, \mathring{H}^{k,\lambda})$ is a surjective operator with nullity $\xi(\infty,k,\lambda)$. By Theorem 1.3.1(i), $A \in L(\mathring{W}_{\infty,l}^{k,\lambda}, \mathring{W}_{\infty}^{k,\lambda})$ is a surjective operator with kernel index $\xi(\infty,k,\lambda)$. So it remains to verify that $y \in \mathring{W}_{\infty,l}^{k,\lambda}$ and $Ay \in \mathring{H}^{k,\lambda}$ imply $y \in \mathring{H}_l^{k,\lambda}$.

It is easily seen that $y \in \mathring{C}_l^k$. We show that the number

$$M = \sup_{0 \le x < t \le b} |y^{(k)}(x) - y^{(k)}(t)|\,|x-t|^{-\lambda}$$

is finite. For $x = 0$ and $0 < t \le b$, we get

$$|y^{(k)}(t)|\, t^{-\lambda} \le \|y^{(k)}\|_{L^{\lambda}_{\infty}} < \infty \quad .$$

Furthermore, for $x > 0$ and $t - x \le x$, with some $\xi \in (x,t)$ we obtain

$$|y^{(k)}(t) - y^{(k)}(x)|\, |t - x|^{-\lambda} \le |y^{(k+1)}(\xi)|\, |t - x|^{1-\lambda}$$

$$\le |y^{(k+1)}(\xi)|\, x^{1-\lambda} \le \xi^{-\lambda} |\xi\, y^{(k+1)}(\xi)| \le \|xy^{(k+1)}\|_{L^{\lambda}_{\infty}} < \infty \, .$$

Finally, for $t - x > x > 0$, we find that

$$|y^{(k)}(t) - y^{(k)}(x)|\, |t - x|^{-\lambda} \le |y^{(k)}(x)|\, |t - x|^{-\lambda} + |y^{(k)}(t)|\, |t-x|^{-\lambda}$$

$$\le |y^{(k)}(x)|\, x^{-\lambda} + 2^{\lambda}\, |y^{(k)}(t)|\, t^{-\lambda} \le (1+2^{\lambda})\, \|y^{(k)}\|_{L^{\lambda}_{\infty}} < \infty \, ,$$

since $t - x > t/2$. Combining the last three estimates, we conclude $M < \infty$, hence $y \in \mathring{H}^{k,\lambda}$ (cf. (1.1.2)). Similarly, $x^i D^i y \in \mathring{H}^{k,\lambda}$ for $i=1,\dots,l-1$. Together with $Ay \in \mathring{H}^{k,\lambda}$, this gives $y \in \mathring{H}^{k,\lambda}_{l}$ Q.E.D.

(ii) The arguments are analogous to those in Theorems 1.3.1(ii) and 1.4.2(ii). □

1.6. Basis of solutions for the homogeneous equation

In this section we shall extend Lemma 1.2.1 to the homogeneous Fuchsian equation

$$Ay = [x^l D^l + \sum_{0 \le i < l} a_i(x) x^i D^i]\, y = 0 \, . \tag{1.6.1}$$

As in 1.2.1 let μ_i $(i=1,\dots,\bar{l})$ be the different characteristic roots of A with multiplicities r_i.

Theorem 1.6.1. Assume (1.3.4) for $k = 0$. Then equation (1.6.1) has a fundamental set of solutions of the form

$$y_{ij}(x) = x^{\mu_i}[(\ln x)^j + v_{ij}(x)],\ j=0,\dots,r_i-1,\ i=1,\dots,\bar{l}, \tag{1.6.2}$$

on $(0,b]$, where $v_{ij} \in \mathring{C}^{0,\varrho}_{l}[0,b]$ for all $\varrho < \sigma$.

Proof. Fix i and consider the operators

$$A_i = x^{-\mu_i} A x^{\mu_i},\ A_{0i} = x^{-\mu_i} A_0 x^{\mu_i},\ T_i = A_i - A_{0i} \, ,$$

where A_0 is the corresponding Euler operator (1.0.2). By assumption, the coefficients of T_i belong to $\mathring{C}^{0,\sigma}[0,b]$. Hence $T_i(\ln x)^j \in \mathring{C}^{0,\varrho}[0,b]$

for all $\rho<\sigma$ and $j\in\mathbb{N}$. It follows from Theorem 1.3.2 that the equation $A_i v = -T_i(\ln x)^j$ has a solution $v_{ij}\in\bigcap_{\rho<\sigma}\overset{\circ}{C}{}_1^{0,\rho}[0,b]$, since $A_i:\overset{\circ}{C}{}_1^{0,\rho}\to\overset{\circ}{C}{}^{0,\bar\rho}$, $\rho<\bar\rho<\sigma$, is surjective and has the same kernel index for all ρ and $\bar\rho$ sufficiently close to σ. Furthermore, by Lemmas 1.2.1 and 1.2.2, we have $A_{oi}(\ln x)^j=0$ $(j=0,\dots,r_i-1)$ on $(0,b]$. Thus, for $j=0,\dots,r_i-1$, y_{ij} defined in (1.6.2) is a solution of (1.6.1) on $(0,b]$. Since the functions y_{ij} are linearly independent, we obtain the result. □

Under stronger assumptions on the smoothness of the coefficients, one can describe the asymptotic behaviour of the basis of solutions more precisely. For fixed $i\in\{1,\dots,l\}$ and $m\in\mathbb{N}$, let z_{im} be the number of characteristic roots $\mu_n(n=1,\dots,l)$ satisfying $\mu_n-\mu_i\in\{1,\dots,m\}$.

Theorem 1.6.2. Assume $k\in\mathbb{N}$ and (1.3.4). Then each function $v=v_{ij}$ in (1.6.2) has the representation

$$v(x)=\sum_{1\le m\le k}\ \sum_{0\le n\le q_m} c_{mn}x^m(\ln x)^n + v_k(x) \qquad (1.6.3)$$

where $c_{mn}\in\mathbb{C}$, $q_m\le j+z_m$ and $v_k\in\overset{\circ}{C}{}_1^{k,\rho}[0,b]$ for all $\rho<\min(\sigma,1)$.

Proof. Fixing i and j and using the notation introduced in the proof of the preceding theorem, we proceed by induction on k. By Theorem 1.6.1, (1.6.3) holds for $k=0$. Suppose the assertion is true for all $k\le\bar k$, where $\bar k\ge 1$, and assume (1.3.4) with $k=\bar k+1$. Then

$$g := A_i\left[(\ln x)^j + v-v_{\bar k}\right] = -A_i v_{\bar k}\in\overset{\circ}{C}{}^{\bar k,\rho}[0,b]$$

for all $\rho<\min(\sigma,1)$, and by (1.3.4) for $k=\bar k+1$, g can be written $g = g_0+g_1$, where $g_0\in\overset{\circ}{C}{}^{\bar k+1,\rho}[0,b]$ $(\rho<\min(\sigma,1))$ and

$$g_1(x)=\sum_{0\le n\le q_{\bar k}} d_n x^{\bar k+1}(\ln x)^n,\ d_n\in\mathbb{C},\ q_{\bar k}\le j+z_{i\bar k}.$$

Furthermore, the Euler equation $A_{oi}z=g_1$ has a solution

$$z_1(x)=\sum_{0\le n\le q_{\bar k+1}} b_n x^{\bar k+1}(\ln x)^n, b_n\in\mathbb{C}, q_{\bar k+1}\le j+z_{i\bar k+1}.$$

Since A_{oi} has the characteristic roots $\mu_n-\mu_i$ $(n=1,\dots,l)$, this follows by successive applications of the following fact: For any $\mu\in\mathbb{C}$ and $j,k\in\mathbb{N}_0$, the equation $(xD-\mu)y = x^k(\ln x)^j$ has the solution

$$y(x) = \begin{cases} x^k(\ln x)^{j+1}/(j+1) & \text{when } \mu = k, \\ \sum_{0 \le i \le j} (-1)^i x^k (xD)^i (\ln x)^j/(k-\mu)^{i+1} & \text{when } \mu \neq k, \end{cases}$$

on $(0,b]$. On the other hand, since $g_0 - T_i z_1 \in \mathring{C}^{\bar{k}+1,\varrho}$ ($\varrho < \min(\sigma,1)$), Theorem 1.3.2 implies that the equation $A_i y = g_0 - T_i z_1$ has a solution $v_{\bar{k}+1} \in \bigcap_{\varrho<1,\sigma} \mathring{C}_1^{\bar{k}+1,\varrho}[0,b]$. Then $(\ln x)^j + (v - v_{\bar{k}} - z_1) - v_{\bar{k}+1}$ is a solution of $A_i y = 0$ on $(0,b]$ which proves (1.6.3) for $k = \bar{k}+1$. □

Under certain additional assumptions on the characteristic roots of A, the logarithmic terms disappear in (1.6.3).

<u>Theorem 1.6.3.</u> Assume (1.3.1), $k \in \mathbb{N}$ and, for $i,j=1,\dots,l$,

$$\mu_i - \mu_j \neq 0,1,\dots,k-1 \quad \text{when } i \neq j, \tag{1.6.4}$$

$$\operatorname{Re}(\mu_i - \mu_j) \neq k. \tag{1.6.5}$$

Then (1.6.1) has a basis of solutions of the form

$$y_i(x) = x^{\mu_i} z_i(x), z_i \in C_1^k[0,b], z_i(0) = 1 \quad (i=1,\dots,l).$$

<u>Proof.</u> It suffices to show that, with the above notation, each equation $A_i z = 0$ admits a solution $z \in C_1^k[0,b]$ satisfying $z(0) = 1$. A_i has the characteristic roots $\mu_n - \mu_i$, $n=1,\dots,l$ (cf. (1.2.3)). Therefore, by (1.6.5), A_i satisfies condition (1.4.10) with $\varrho = 0$. Moreover, in view of (1.6.4) there exists a polynomial β of degree k satisfying $\beta(0) = 1$ and $A_i\beta \in \mathring{C}^k[0,b]$; see 1.4.1. Consequently, by Theorem 1.3.1(i), we can find an element $y \in \mathring{C}_1^k[0,b]$ such that $A_i(\beta - y) = 0$.

Q.E.D. □

Finally, we generalize Theorem 1.6.2 to the case when A has C^∞ coefficients. For fixed $i \in \{1,\dots,\bar{l}\}$, let z_i be the number of characteristic roots $\mu_n (n=1,\dots,l)$ satisfying $\mu_n - \mu_i \in \mathbb{N}$.

<u>Theorem 1.6.4.</u> Under hypothesis (1.5.1), each function $v=v_{ij}$ in (1.6.2) has the representation

$$v(x) = \sum_{0 \le n \le s} \varphi_n(x)(\ln x)^n, \quad s \le j + z_i,$$

where $\varphi_n \in C^\infty[0,b]$ and $\varphi_n(0) = 0$ for all n.

<u>Proof.</u> Fix i and j and define A_i as above. Replacing the coefficients of A_i by their formal Taylor series at $x=0$, we obtain a formal diffe-

rential operator $\hat{A}_1$. Taking $k \to \infty$ in (1.6.3), we get a formal log-power series

$$\hat{v}(x) = (\ln x)^j + \sum_{0 \leq n \leq s} \sum_{m \geq 1} c_{nm} x^m (\ln x)^n, \quad c_{mn} \in \mathbb{C}, \quad s \leq j + z_1,$$

such that $\hat{A}_1 \hat{v} = 0$. By Borel's theorem (cf. Narasimhan [1, Chap. 1]), there exist functions $\psi_n \in C^\infty[0,b]$ such that $\psi_n^{(m)}(0)/m! = c_{nm} (m \in \mathbb{N})$ and $\psi_n(0) = 0$. Consequently, the function

$$g := A_1 [(\ln x)^j + \sum_{0 \leq n \leq s} \psi_n(x)(\ln x)^n]$$

belongs to $C^\infty[0,b]$ and $g^{(m)}(0) = 0$ for all $m \in \mathbb{N}_0$. Furthermore, by Theorem 1.5.2 applied to the Fuchsian operator $x^{-1} A_1 x$, there exists $z \in C^\infty[0,b]$ such that $A_1 z = g$ and $z(0) = 0$. Setting

$$v(x) = \sum_{0 \leq n \leq s} \psi_n(x)(\ln x)^n - z(x),$$

we thus obtain $A_1 [(\ln x)^j + v(x)] = 0$ on $(0,b]$ Q.E.D. □

1.7. Maximal Fuchsian operators

1.7.1. For $k \in \mathbb{N}$, define $\mathcal{A}^k(0,b)$ to be the set of functions y on $(0,b]$ for which $y^{(k-1)}$ exists and is absolutely continuous on every closed subinterval of $(0,b]$. Let A be the Fuchsian operator (1.o.1). In this section we consider the maximal Fuchsian operator $A_{p,k}^{\varrho,\bar{\varrho}}$ and $A_k^{\varrho,\bar{\varrho}}$, $\bar{\varrho} \geq \varrho$, defined as follows:

$$D(A_{p,k}^{\varrho,\bar{\varrho}}) = \{y \in W_p^{k,\varrho}(0,b) \cap \mathcal{A}^{1+k}(0,b) : Ay \in W_p^{k,\bar{\varrho}}(0,b)\}, \; A_{p,k}^{\varrho,\bar{\varrho}} y = Ay;$$

$$D(A_k^{\varrho,\bar{\varrho}}) = \{y \in C^{k,\varrho}[0,b] \cap \mathcal{A}^{1+k}(0,b) : Ay \in C^{k,\bar{\varrho}}[0,b]\}, \; A_k^{\varrho,\bar{\varrho}} y = Ay.$$

The following theorem shows that, under rather weak smoothness assumptions on the coefficients, the maximal Fuchsian operators coincide with the operators studied in 1.4.3 and 1.4.4. For $\bar{\varrho} > \varrho$, let $D_{p,k}^{\varrho,\bar{\varrho}}(A)$ and $D_k^{\varrho,\bar{\varrho}}(A)$ be the domains of definition (1.4.11) and (1.4.12), respectively. Let $D_{p,k}^{\varrho,\varrho}(A) = W_{p,1}^{k,\varrho}(0,b)$ and $D_k^{\varrho,\varrho}(A) = C_1^{k,\varrho}[0,b]$.

Theorem 1.7.1. Assume (1.3.4) and $\bar{\varrho} \geq \varrho$. If ϱ satisfies condition (1.4.5) and (1.4.9), then $D(A_{p,k}^{\varrho,\bar{\varrho}}) = D_{p,k}^{\varrho,\bar{\varrho}}(A)$ and $D(A_k^{\varrho,\bar{\varrho}}) = D_k^{\varrho,\bar{\varrho}}(A)$, respectively.

Proof. 1. Let us show the second assertion. The proof of the first one is analogous. It is sufficient to consider the case $\bar{\varrho} = \varrho$. Indeed, since $C^{k,\bar{\varrho}} \subset C^{k,\varrho}$ for $\bar{\varrho} > \varrho$, $y \in D(A_k^{\varrho,\bar{\varrho}}) \subset D(A_k^{\varrho,\varrho})$ and $D(A_k^{\varrho,\varrho}) = C_l^{k,\varrho}$ would imply $y \in D_k^{\varrho,\bar{\varrho}}(A)$. Furthermore, the proof reduces to the case k=0: For j=1,...,k and $k \in \mathbb{N}$, we have $D^j A = A_j D^j + B_{j-1}$, where A_j is a Fuchsian operator of order l and B_{j-1} a differential operator of order $< j$ with coefficients in $C^{0,\sigma}$, $\sigma > 0$. Suppose that $y \in D(A_k^{\varrho,\varrho})$ and Theorem 1.7.1 holds for k=0. Then $D^j y \in D((A_j)_0^{\varrho,\varrho})$, and $D((A_j)_0^{\varrho,\varrho}) = C_l^{0,\varrho}$ would imply $D^j y \in C_l^{0,\varrho}$, j=1,...,k.

Thus it remains to prove that, under hypothesis (1.3.4) for k=0, $y \in D(A_0^{\varrho,\varrho})$ implies $y \in C_l^{0,\varrho}[0,b]$.

2. We now verify the last assertion under the additional hypotheses

$$\mu_i \neq 0 \ , \ \operatorname{Re}\mu_i \neq \varrho, \ i=1,\dots,l \ , \tag{1.7.1}$$

$$\operatorname{Re}(\mu_i - \mu_j) \neq 0 \quad \text{for all } i \neq j. \tag{1.7.2}$$

By (1.7.1) and Theorem 1.4.4(i), for any $y \in D(A_0^{\varrho,\varrho})$, there exists $\bar{y} \in C_l^{0,\varrho}[0,b]$ such that $A(\bar{y}-y) = 0$. Therefore it suffices to check the relation

$$v \in D(A_0^{\varrho,\varrho}) \quad \text{and } Av = 0 \Rightarrow v \in C_l^{0,\varrho}[0,b] \ . \tag{1.7.3}$$

By (1.7.2) and Theorem 1.6.1, v can be written

$$v(x) = \sum_{1 \le i \le l} c_i z_i(x) x^{\mu_i} \ , \ c_i \in \mathbb{C} \ , \ z_i \in C_l^0[0,b], \ z_i(0) = 1 \quad (i=1,\dots,l) \ .$$

In view of (1.7.1) and (1.7.2) it is clear that $v \in C^{0,\varrho}[0,b]$ if and only if $c_i = 0$ for all i satisfying $\operatorname{Re}\mu_i < \varrho$. Hence $v \in C_l^{0,\varrho}$ and (1.7.3) is proved.

3. If one of the conditions (1.7.1) and (1.7.2) is violated, then we pass to the operators $A_\varepsilon = A - \varepsilon$ and show that A_ε satisfies (1.7.1) and (1.7.2) for some $\varepsilon \neq 0$. The characteristic equation of A_ε is

$$P(\mu) - \varepsilon = 0 \ , \tag{1.7.4}$$

where P is the characteristic polynomial (1.2.1) of A. We choose a complex number c such that $|c| = 1$, $\operatorname{Re} c\,\eta_i \neq 0$ (i=1,...,l) and $\operatorname{Re} c(\eta_i - \eta_j) \neq 0$ for all $i \neq j$, where η_i are the roots of the equation $\eta^l = 1$. Setting $z = \mu / c\,\varepsilon_1^{1/l}$ and $\varepsilon = \varepsilon_1 c^l$, $\varepsilon_1 > 0$, (1.7.4) becomes

$$Q_\varepsilon(z) := \varepsilon[z^l + \sum_{0 \le i \le l} b_i \varepsilon_1^{i/l-1} c^{i-l} z^i - 1] = 0$$

with some $b_i \in \mathbb{C}$. We obviously have $\varepsilon_1^{i/l-1} \to 0$ $(i=0,\dots,l-1)$ as $\varepsilon_1 \to \infty$. Since the roots of an algebraic equation depend continuously on the coefficients, the characteristic roots of A_ε satisfy (1.7.1) and (1.7.2) when ε_1 is large enough. (Note that $\varepsilon^{-1} Q_\varepsilon(z) \to z^l - 1$ as $\varepsilon_1 \to \infty$) Now $y \in D(A_0^{\varrho,\varrho})$ implies $y \in D((A_\varepsilon)_0^{\varrho,\varrho})$, hence $y \in C^{o;\varrho}_{l}$. □

<u>1.7.2.</u> As an application of the results in 1.4.4 and 1.7.1, we study the following more general degenerate differential operator:

$$B = x^{l-r} D^l + \sum_{0 \le i < l} a_i(x) x^{i-r} D^i \quad , \qquad r > 0 .$$

With B we associate the Fuchsian operator

$$A = x^r B = x^l D^l + \sum_{0 \le i < l} a_i(x) x^i D^i$$

and the characteristic equation (1.2.1). Consider B as a linear operator in $L_p(0,b)$, $1 \le p \le \infty$, with domain of definition

$$D_p(B) = \{ y \in L_p(0,b) \cap \mathcal{A}^l(0,b) : By \in L_p(0,b) \} . \qquad (1.7.5)$$

<u>Theorem 1.7.2.</u> Assume (1.3.4) for k=0. Then the operator $B : L_p(0,b) \to L_p(0,b)$ with domain of definition (1.7.5) is surjective and has kernel index $\xi(p,0,0) + \xi^*(p,0,0)$. Moreover,

$$D_p(B) = \{ y \in W^o_{p,l}(0,b) : By \in L_p(0,b) \} . \qquad (1.7.6)$$

<u>Proof.</u> Consider the maximal Fuchsian operator $A^{o,r}_{p,o} : L_p(0,b) \to L^r_p(0,b)$ which corresponds to A. Then (1.7.6) follows from Theorem 1.7.1 and the first assertion is then a consequence of Theorem 1.4.6. □

<u>1.8. Comments and references</u>

<u>1.2.</u> The proof of Lemma 1.2.5 is due to Elschner and Silbermann [1]. Another method of proof can be found in Gluško [1]. The Euler operator in weighted Sobolev spaces has been studied in several papers. See for example Bolley and Camus [1], Bolley, Camus and Helffer [1], Višik and Grušin [1]. The normal solvability of the Euler operator in $L_p(1,\infty)$ was investigated in Goldberg [1, Chap. 6], while Müller-Pfeiffer [1] studied this operator in $L_2(0,\infty)$ by Hilbert space methods.

1.4. Under the hypothesis $\operatorname{Re}\mu_i \bar{\in} [-1/p, -1/p+k]$ (resp. $\operatorname{Re}\mu_i \bar{\in} [0,k]$) $(i=1,\ldots,l)$, Theorem 1.4.2(i) (resp. 1.4.4(i)) for $\varrho=0$ was proved by other methods in Gluško [2]. In these cases no solvability conditions occur. For $\varrho=0$ and $\operatorname{Re}\mu_i < k$ $(i=1,\ldots,l)$, Theorem 1.4.4(i) is also contained in the results of Baouendi and Goulaouic [1]. Elschner and Silbermann [1] computed the index of the Fuchsian operator in the general case. In this paper, the Fuchsian operator was considered as a compact perturbation of the corresponding Euler operator. Bolley, Camus and Helffer [1], [2] used similar methods in order to compute the index of Fuchsian operators with polynomial coefficients. Furthermore, several authors studied systems of linear differential equations of the form $xY'(x) + A(x)Y(x) = F(x)$ in the spaces W_p^k and C^k of vector-valued functions (cf. Gluško [2], Natterer [1], de Hoog and Weiss [1], Elschner [2]). These results can be deduced and generalized using the methods of this section.

1.5. Theorem 1.5.1 was proved in Elschner and Silbermann [1]. By other methods, Theorem 1.5.2 in a somewhat less precise form was derived in Elschner [1]. Under the hypothesis $\operatorname{Re}\mu_i \bar{\in} [0,k+\lambda]$ $(i=1,\ldots,l)$, Theorem 1.5.4(i) was obtained in Gluško [2].

1.6. Under the condition $\operatorname{Re}\mu_i = \operatorname{Re}\mu_j \Rightarrow \mu_i = \mu_j$ $(i,j=1,\ldots,l)$, Theorem 1.6.1 was shown by Natterer [2] using other methods. Theorem 1.6.2 may be considered as a generalization of well-knowns results concerning the asymptotic behaviour of solutions to homogeneous Fuchsian equations with analytic coefficients (see Wasow [1], Coddington and Levinson [1]). Theorem 1.6.3 was already proved by Lomov [1].

1.7. For general properties of maximal operators corresponding to linear differential operators in L_p spaces, we refer to Goldberg [1]. Under somewhat stronger smoothness assumptions on the coefficients, Theorem 1.7.1 (for the C^k spaces) was shown in Baouendi and Goulaouic [1]. For $r=1$, Theorem 1.7.2 can be found in Natterer [2].

2. LINEAR ORDINARY DIFFERENTIAL OPERATORS WITH ONE SINGULAR POINT

In this chapter we consider differential operators of the form

$$A = \sum_{0 \le i \le l} a_i(x) D^i \ , \ a_i \in C^{\infty}[a,b] \quad (i=0,\dots,l) \ , \ 0 \in [a,b],$$
$$a_l(x) = x^q \bar{a}_l(x), \ q \in \mathbb{N}, \quad \bar{a}_l(x) \neq 0 \quad \text{for all } x \in [a,b], \tag{2.o.1}$$

having a singularity at the origin. We mainly study (2.o.1) as an operator acting from $L_p^{\varrho}(0,b)$ into $L_p^{\bar{\varrho}}(0,b)$, $\bar{\varrho} \ge \varrho$, and as a mapping in $C^{\infty}[a,b]$. It turns out that A is always a Fredholm operator in C^{∞}, and a formula for the index will be given. However, in contrast to Fuchsian operators, $A : L_p^{\varrho} \to L_p^{\bar{\varrho}}$ is normally solvable only when the number $\bar{\varrho} - \varrho$ is sufficiently large. Following the general outline of Chap. 1, we choose a principal part A_o of (2.o.1) and appropriate weighted Sobolev spaces such that A becomes a "small" perturbation of A_o in those spaces and the index of A_o may be computed. The corresponding constructions, however, depend on the factorization of formal differential operators and are much more complicated than in the case of Fuchsian operators.

2.1. Factorization of formal differential operators

2.1.1. An expression of the form

$$c(x) = \sum_{j \ge M} c_j x^{j/m} \ , \ c_j \in \mathbb{C} \ , \ M \in \mathbb{Z} \ , \ m \in \mathbb{N}, \tag{2.1.1}$$

is called a formal Puiseux series, where $x^{1/m}$ denotes a fixed continuous branch of this function in $\mathbb{R} \setminus \{0\}$ such that $x^{1/m} = |x|^{1/m}$ for $x > 0$. In the following we set $v(c) = M/m$ when $c_M \neq 0$ and $v(c) = \infty$ when $c = 0$, i.e. $c_j = 0$ for all $j \ge M$.

For fixed m, let $\mathfrak{A}_m$ be the set of formal series (2.1.1) with $M \ge 0$. By $\mathfrak{A}$ we denote the set of all formal Puiseux series (2.1.1) satisfying $M \ge 0$. $\mathfrak{A}_m$ and $\mathfrak{A}$ are rings with respect to the usual multiplication of formal power series. By $\hat{\mathfrak{A}}$ we denote the set of all formal series (2.1.1), which is the field of fractions of the ring $\mathfrak{A}$. Analogously, the set $\hat{\mathfrak{A}}_m$ of all Puiseux series (2.1.1) with fixed $m \in \mathbb{N}$ is the field of fractions of $\mathfrak{A}_m$ (cf. Bourbaki [1, Chap. IV, § 5]).

Let $\hat{\mathfrak{a}}(D)$ be the ring of all formal differential operators (differential polynomials)

$$\hat{A} = \sum_{0 \le i \le l} \hat{a}_i(x) D^i , \quad \hat{a}_i \in \hat{\mathfrak{a}} \quad (i=0,\ldots,l) . \tag{2.1.2}$$

The multiplication in $\hat{\mathfrak{a}}(D)$ is defined by the commutation formulas

$$DcD^i - cD^{i+1} = (Dc)D^i$$

with c given by (2.1.1) and

$$(Dc)(x) = \sum_{j \ge M} (j/m) c_j \, x^{j/m-1} .$$

We say that the differential polynomial (2.1.2) has degree (or order) $l \in \mathbb{N}_0$ if $\hat{a}_l \neq 0$ and degree -1 if $\hat{A}=0$. Since $\hat{\mathfrak{a}}$ is a field, it is not difficult to check that the ring $\hat{\mathfrak{a}}(D)$ is Euclidean, i.e. for any $A,B \in \hat{\mathfrak{a}}(D)$, $B \neq 0$, there exist unique elements $E_1,E_2,F_1,F_2 \in \hat{\mathfrak{a}}(D)$ such that the degree of F_1 and F_2 is smaller than that of B and

$$A = E_1 B + F_1 = BE_2 + F_2 .$$

In order to classify the formal differential operators (2.1.2), we define the characteristic index $\chi(\hat{A})$ of $\hat{A}$ to be $l-l_0$, where the integer $l_0 = l_0(\hat{A})$ $(0 \le l_0 \le l)$ is specified by the requirements

$$\begin{aligned} i-v(\hat{a}_i) &\le l_0 - v(\hat{a}_{l_0}) \quad \text{for } i < l_0 , \\ i-v(\hat{a}_i) &< l_0 - v(\hat{a}_{l_0}) \quad \text{for } i > l_0 . \end{aligned} \tag{2.1.3}$$

Furthermore, we define

$$\varkappa = \varkappa(\hat{A}) = l_0 - v(\hat{a}_{l_0}) = \max_{0 \le i \le l} [i - v(\hat{a}_i)] . \tag{2.1.4}$$

The following theorem due to Kuznecov [1] is basic for our further considerations. Let $\partial = xD$, $l \ge 1$ and $\hat{a}_l \neq 0$ in the sequel.

<u>Theorem 2.1.1.</u> Each differential polynomial of degree $l \ge 1$ has a right divisor of the form $\partial - c$, $c \in \hat{\mathfrak{a}}$.

Applying this theorem l times, we obtain

<u>Corollary 2.1.2.</u> For each formal differential operator $\hat{A}$ of order l, there exist $m \in \mathbb{N}$ and $\lambda_i \in \hat{\mathfrak{a}}_m$ $(i=1,\ldots,l)$ such that

$$\hat{A} = \hat{a}_l(x) x^{-l} (\partial - \lambda_l(x)) \ldots (\partial - \lambda_1(x)) . \tag{2.1.5}$$

In the sequel we often use the following notation: For a formal series

(2.1.1) let $\tilde{c}$ be the finite sum of terms of order <0 in c, i.e.

$$\tilde{c}(x) = \sum_{M \le j \le -1} c_j x^{j/m} \quad \text{when } M<0, \quad \tilde{c}_j(x) = 0 \quad \text{when } M \ge 0.$$

Let λ_i be the formal series from the factorization (2.1.5). Then the functions $\tilde{\lambda}_i$ are called characteristic factors of $\hat{A}$. It turns out that the characteristic factors are uniquely determined by the operator $\hat{A}$.

2.1.1. Proof of Theorem 2.1.1

1. Given a differential polynomial (2.1.2) of degree $l>1$, we consider the operator $B = x^{\varkappa}\hat{A}$ with coefficients $b_i = x^{\varkappa}\hat{a}_i$. By definitions (2.1.3) and (2.1.4), we then have $b_i \in \mathfrak{A}_m$ for some $m \in \mathbb{N}$, $v(b_i) \ge i$ for $i<l_o$, $v(b_{l_o}) = l_o$ and $v(b_i)>i$ for $i>l_o$. We set

$$r = \min_{1 \le i \le l} [-1+v(b_i)/i] ,$$

$$l_1 = \max \{ i : 1 \le i \le l, \ v(b_i) = (r+1)i \}$$

and associate the characteristic triplet (m,r,l_1) with B. Our first goal is to prove that, after a finite number of substitutions

$$x^s\partial \to x^s\partial + \gamma, \quad \gamma \in \mathbb{C} ,$$

and s a positive rational number, B becomes a differential polynomial with r=0. B can be written

$$B = \sum_{0 \le i \le l} c_i(x)(x^r\partial)^i , \quad c_i \in \mathfrak{A} .$$

Assume $r>0$. Then $l_o = 0$, $c_o(0) \ne 0$, $c_{l_1}(0) \ne 0$ and $c_i(0) = 0$ for $i>l_1$. With B we further associate the algebraic equation

$$P(\gamma) = \sum_{0 \le i \le l_1} c_i(0)\gamma^i = 0 . \qquad (2.1.6)$$

Let γ_1 be a root of (2.1.6) with multiplicity $\bar{l}$. Substituting $x^r\partial \to x^r\partial + \gamma_1$ in B, we obtain an operator

$$C = \sum_{0 \le i \le l} d_i(x)(x^r\partial)^i , \quad d_i \in \mathfrak{A} ,$$

and it is easy to check that, for any $r \ge 0$,

$$P(\gamma + \gamma_1) = d_o(0) + \ldots + d_{l_1}(0)\gamma^{l_1} ,$$

since the relation

$$(x^r\partial + \gamma_1)^i = \sum_{0 \le j \le i} \binom{i}{j}\gamma_1^{i-j}(x^r\partial)^j + \sum_{k+j<i} c_{kj}x^{r(i-k)}\partial^j$$

holds with some $c_{kj} \in \mathbb{C}$ and $r(i-k) > rj$ when $r > 0$. Note that the last equality is valid with $c_{kj} = 0$ when $r = 0$.

Furthermore, since $\gamma = 0$ is a root of $P(\gamma + \gamma_1) = 0$ and $l_o(C) = 0$, $\varkappa(C) < 0$. Consider the differential polynomial $x^{\varkappa(C)}C$, and let $\bar{b}_i$ be its coefficients and $(\bar{m}, \bar{r}, \bar{l}_1)$ the characteristic triplet. Since $\gamma = 0$ is a root of $P(\gamma + \gamma_1) = 0$ of multiplicity $\bar{l}$, we have $d_{\bar{l}}(0) \neq 0$ which implies

$$v(\bar{b}_{\bar{l}})/\bar{l} = ((r+1)\bar{l} + \varkappa(C))/\bar{l} < ((r+1)i + \varkappa(C))/i \leq v(\bar{b}_i)/i$$

for all $i > \bar{l}$. Hence $\bar{l}_1 \leq \bar{l} \leq l_1$, and by successive application of the above substitution, we can stabilize the numbers l_1. Thus we may assume $\bar{l}_1 = \bar{l} = l_1$. Then $\bar{m} = m$. Indeed, since (2.1.6) has only the root γ_1 in view of $\bar{l} = l_1$, we get

$$c_1(0) = - c_{\bar{l}}(0)\gamma_1^{\bar{l}-1}\,\bar{l} \neq 0 ,$$

hence $v(b_1) = r$ and $x^r \in \mathfrak{A}_m$ which implies $\bar{m} = m$. Moreover,

$$\bar{r} = \min_{1 \leq i \leq l} [-1 + v(\bar{b}_i)/i] \leq v(\bar{b}_{l_1})/l_1 - 1$$

$$= (rl_1 + \varkappa(C))/l_1 < r$$

and after a finite number of steps, r becomes 0 since $rm \in \mathbb{N}_o$.

2. Let now $r = 0$. Then, with the above notation, we have $c_i \in \mathfrak{A}_m$ and $l_1 = l_o$. Choose γ_1 such that $P(\gamma + \gamma_1) = 0$ has the root $\gamma = 0$ but no positive rational roots, and let C be the operator defined in 1. We show that the equation $Cu = 0$ has a nontrivial solution $u \in \mathfrak{A}_m$. In order to do so, we recurrently define a sequence $\{u_n\} \subset \mathfrak{A}_m$ by the requirements

$$u_o = 1,\ u_n = u_{n-1} + \alpha_n x^{n/m} ,\ \alpha_n \in \mathbb{C} ,$$

$$Cu_n \equiv 0 \pmod{x^{(n+1)/m}}$$

for all $n \in \mathbb{N}$. We have

$$Cu_o = d_o \equiv 0 \pmod{x^{1/m}}$$

$$Cu_n = Cu_{n-1} + \alpha_n x^{n/m} \sum_{0 \leq i \leq l_o} d_i(\gamma_1 + n/m)^i,\ n \in \mathbb{N} .$$

Therefore the numbers α_n are uniquely determined since $P(\gamma_1 + n/m) \neq 0$

for all n and m. Setting $u = u_0 + \sum_{n\geq 1} \alpha_n x^{n/m}$, we obtain the desired solution of $Cu = 0$.

Finally, we verify that C has the right divisor $\partial - \partial u/u$, completing the proof of the theorem. Since the ring $\hat{\mathfrak{A}}(D)$ is Euclidean, there exist differential polynomials E and F of degree l-1 and ≤ 0, respectively, such that $C = E(\partial - \partial u/u) + F$. But $Fu = Cu - E(\partial - \partial u/u)u = 0$ and $u \neq 0$ imply $F = 0$. □

2.1.3. We now determine some properties of the factorization (2.1.5) and the characteristic factors.

Lemma 2.1.3. Let $\lambda_i \in \hat{\mathfrak{A}}_m$ (i=1,2), $\tilde{\lambda}_1 \neq \tilde{\lambda}_2$ and $r_0 = v(\lambda_1 - \lambda_2)$. Then the commutation formula

$$(\partial - \lambda_1)(\partial - \lambda_2) = (\partial - \lambda_2 - r)(\partial - \lambda_1 + r) \tag{2.1.7}$$

holds with some $r \in \mathfrak{A}_m$ such that $r(0) = r_0$.

Proof. (2.1.7) can be written

$$r + (\partial r - r^2)/(\lambda_1 - \lambda_2) = (\partial\lambda_1 - \partial\lambda_2)/(\lambda_1 - \lambda_2). \tag{2.1.8}$$

Set $r = r_0 + r_1 x^{1/m} + \ldots$. Since

$$(\partial\lambda_1 - \partial\lambda_2)/(\lambda_1 - \lambda_2) = v(\lambda_1 - \lambda_2) + \sum_{j\geq 1} d_j x^{j/m}, \quad d_j \in \mathbb{C}$$

and $v((\partial r - r^2)/(\lambda_1 - \lambda_2)) > 0$, we may uniquely determine the numbers $r_0, r_1, \ldots$ such that (2.1.8) holds. □

Lemma 2.1.4. Let $\lambda_i \in \hat{\mathfrak{A}}$ (i=1,...,l). Then

$$\chi[(\partial - \lambda_l)\ldots(\partial - \lambda_1)] = \chi(\partial - \lambda_1) + \ldots + \chi(\partial - \lambda_l). \tag{2.1.9}$$

Proof. We proceed by induction on l. First, we observe that $\chi(\partial - \lambda) = 0$ for $\lambda \in \hat{\mathfrak{A}}$ if and only if $\tilde{\lambda} = 0$, and $\chi(\partial - \lambda) = 1$ otherwise. Assume now that formula (2.1.9) holds for l=k, k > 1. Consider the operator $B = (\partial - \lambda)A$, where $\lambda \in \hat{\mathfrak{A}}$ and

$$A = (\partial-\lambda_k)\dots(\partial-\lambda_1) = x^kD^k + \sum_{0\le i<k} a_iD^i\ , \quad a_i \in \hat{\mathfrak{A}}\ .$$

Then we have

$$B = x^{k+1}D^{k+1} + \sum_{0\le i\le k} b_iD^i\ , \quad b_i = -\lambda a_i + xa_{i-1} + \partial a_i \quad (i=0,\dots,k)$$

with the convention $a_{-1} = 0$, $a_k = x^k$. Let l_0 be the number (2.1.3) which corresponds to A. Let $\tilde{\lambda} = 0$. Since $v(a_{l_0+1}) > v(a_{l_0}) + 1$,

$$l_0-v(a_{l_0}) = l_0+1-v(b_{l_0+1})\ .$$

Moreover,

$$k+1-v(x^{k+1}) = k-v(x^k) = 0\ ,$$

$$i-v(b_i) \le \max\{i-v(a_i), i-1-v(a_{i-1})\}\ , \quad i=0,\dots,k\ ,$$

which implies $\chi(B) = l+1-(l_0+1) = \chi(A)$.

Let now $\tilde{\lambda} \neq 0$. Then $v(\lambda) < 0$ and

$$l_0-v(b_{l_0}) = l_0-v(a_{l_0})-v(\lambda)\ ,$$

$$i-v(b_i) \le \max\{i-v(a_i)-v(\lambda), i-1-v(a_{i-1})\}\ , \quad i=0,\dots,k\ .$$

Hence $\chi(B) = l+1-l_0 = \chi(A)+1$, and the assertion is proved for $l = k+1$. □

Corollary 2.1.5. In each factorization of the form (2.1.5) exactly l_0 characteristic factors vanish identically.

Lemma 2.1.6. The characteristic factors of a formal differential operator (2.1.2) are uniquely determined by its coefficients.

Proof. Consider a factorization (2.1.5) of $\hat{A}$, and let $\tilde{\lambda}_i$ be the corresponding characteristic factors. For $i=1,\dots,l$, define

$$\hat{\lambda}_i = \int \tilde{\lambda}_i x^{-1}dx \qquad (2.1.10)$$

with the convention

$$\int\Big(\sum_{j\neq 0} c_jx^{j/m}\Big)x^{-1}dx = \sum_{j\neq 0} (m/j)c_jx^{j/m}\ , \quad c_j \in \mathbb{C}\ .$$

Let $\tilde{\lambda}_1$, for example, be a characteristic factor of multiplicity l_1. By Lemma 2.1.4,

$$\chi(e^{-\hat{\lambda}_1}\hat{A}e^{\hat{\lambda}_1}) = \chi[(\partial-\lambda_l+\tilde{\lambda}_1)\dots(\partial-\lambda_1+\tilde{\lambda}_1)] = l-l_1. \quad (2.1.11)$$

Consider another factorization

$$\hat{A} = \hat{a}_l x^{-1}(\partial - \Lambda_l)\dots(\partial - \Lambda_1) \quad , \quad \Lambda_i \in \hat{\mathfrak{A}} \, .$$

Then (2.1.11) implies

$$\chi\,[(\partial - \Lambda_l + \tilde{\lambda}_1)\dots(\partial - \Lambda_1 + \tilde{\lambda}_1)] = l - l_1 \, ,$$

and by Lemma 2.1.4, there exist exactly l_1 indices i such that $\tilde{\Lambda}_i = \tilde{\lambda}_i$. □

Remark 2.1.7. Let $\lambda \in \hat{\mathfrak{A}}$ and $\hat{\lambda} = \int \tilde{\lambda} x^{-1} dx$. It follows from the above proof that $\chi(e^{-\hat{\lambda}} \hat{A} e^{\hat{\lambda}}) = l-k$ if and only if $\tilde{\lambda}$ coincides with a characteristic factor of $\hat{A}$ of multiplicity k.

Therefore the functions $\hat{\lambda}_i (i=1,\dots,l)$ defined in (2.1.10) coincide with the determining factors of $\hat{A}$ which were used in Kannai [2] and Helffer and Kannai [1].

It follows from Lemma 2.1.3 that the complete formal series λ_i in (2.1.5) are not unique, in general. The next lemma and the remark following it, however, show that the zero order terms of λ_i are unique when the order of the characteristic factors $\tilde{\lambda}_i$ is fixed in (2.1.5).

Lemma 2.1.8. Let $\hat{A}$ be the operator (2.1.2) with factorization (2.1.5), $l_0 > 0$ and

$$\tilde{\lambda}_i = 0 \ (i=1,\dots,l_0) \, , \quad \tilde{\lambda}_i \neq 0 \ (i=l_0+1,\dots,l) \, .$$

Define $\mu_i = (\lambda_i - \tilde{\lambda}_i)(0)$. Then the numbers $\mu_1,\dots,\mu_{l_0}$ coincide with the roots of the characteristic equation

$$\begin{aligned} P(\mu) = x^{-\mu+\varkappa} \hat{A}(x^{\mu})|_{x=0} &= b_{l_0}(0)\mu(\mu-1)\dots(\mu-l_0+1) \\ &+ b_{l_0-1}(0)\mu\dots(\mu-l_0+2)+\dots+b_1(0)\mu+b_0(0) = 0 \end{aligned} \tag{2.1.12}$$

of $\hat{A}$, where $b_i = x^{\varkappa-i}\hat{a}_i$.

Proof. By assumption and definitions (2.1.3) and (2.1.4), it follows from (2.1.5) that

$$\begin{aligned} P(\mu) &= x^{-\mu+\varkappa} \hat{a}_l x^{-1}[(\partial - \lambda_l)\dots(\partial - \lambda_1)](x^{\mu})|_{x=0} \\ &= x^{-\mu+\varkappa} \hat{a}_l x^{-1}(-1)^{l-l_0} \lambda_l \dots \lambda_{l_0+1} [(\partial-\lambda_{l_0})\dots(\partial-\lambda_1)](x^{\mu})|_{x=0} \\ &= b_{l_0}(0) x^{-\mu}[(\partial-\mu_{l_0})\dots(\partial-\mu_1)](x^{\mu}) = b_{l_0}(0)(\mu-\mu_{l_0})\dots(\mu-\mu_1). \ \square \end{aligned}$$

Remark 2.1.9. Let $\tilde{\chi}$ be a characteristic factor of A of multiplicity k. By Lemma 2.1.3, the factors in (2.1.5) can be labelled in such a way that $\tilde{\chi}_i = \tilde{\chi}$ $(i=1,\ldots,k)$. Consider the operator

$$A_1 = e^{-\hat{\chi}}\hat{A}e^{\hat{\chi}} \in \hat{\mathcal{A}}(D) , \quad \hat{\chi} = \int \tilde{\chi}\, x^{-1}dx.$$

Then $\chi(A_1) = 1-k$ (cf. the proof of Lemma 2.1.6) and Lemma 2.1.8 implies that the numbers $\mu_1,\ldots,\mu_k$ defined above coincide with the roots of the characteristic equation of A_1. Using this argument for all different characteristic factors of A and relation (2.1.7), we find that the numbers μ_i $(i=1,\ldots,l)$ are unique when the order of the characteristic factors is fixed in (2.1.5).

2.2. On the computation of characteristic factors

2.2.1. For a non-vanishing characteristic factor $\tilde{\chi}_i$ of (2.1.2), let $\gamma_i x^{1-q_i}$ be its lowest term, where $q_i = 1-v(\tilde{\chi}_i)$. By Lemma 2.1.3 there exists a factorization (2.1.5) such that

$$1 < q_i \le q_{i+1} \quad \text{for } i=l_0+1,\ldots,l-1 . \tag{2.2.1}$$

We shall assume from now on, without explicitely mentioning it, that (2.2.1) holds. Then it is not difficult to compute the lowest terms of the characteristic factors $\tilde{\chi}_i$ $(i>l_0)$.

Lemma 2.2.1. Let $l_0 < l$. Then for some index s $(1\le s\le l)$,

$$q_i = \min_{k\ge l(j)+1} \frac{v(\hat{a}_k)-v(\hat{a}_{l(j)})}{k-l(j)} , \; i=l(j)+1,\ldots,l(j+1) , \tag{2.2.2}$$

where

$$l(j+1) = \max\{\, i : l(j)+1\le i\le l, v(\hat{a}_i)-v(\hat{a}_{l(j)}) = q_{l(j)+1}(i-l(j))\}$$

for $j=0,\ldots,s-1$ and $l(0)=l_0$, $l(s)=l$.

Moreover, the numbers γ_i $(i=l(j)+1,\ldots,l(j+1))$ coincide with the roots of the algebraic equation

$$\sum_{0\le i\le l(j+1)-l(j)} b_{l(j)+i}(0)\gamma^i = 0 , \tag{2.2.3}$$

where

$$b_{l(j)+i} = x^{-\varkappa(i,j)}\hat{a}_{l(j)+i} , \quad \varkappa(i,j) = v(\hat{a}_{l(j)}) + i q_{l(j+1)} .$$

Proof. Let $q_i (i > l_0)$ and $l(j)$ be the numbers defined in (2.2.2), $q_0=0$, $q_i=1$ $(i=1,\dots,l_0)$ and

$$q(i) = q_0 + \dots + q_i \quad (i=0,\dots,l) . \tag{2.2.4}$$

Then the operator (2.1.2) can be written

$$\hat{A} = \sum_{0 \le i \le l} b_i x^{q(i)-\varkappa} D^i, \; b_i \in \mathfrak{A}, b_{l(j)}(0) \neq 0 (j=0,\dots,s), \tag{2.2.5}$$

where b_i $(i \ge l(0))$ are defined as above. Furthermore, from (2.1.5) and the commutation relation

$$(\partial - a)x^{\varrho} = x^{\varrho} (\partial - a + \varrho), \; a \in \hat{\mathfrak{A}}, \quad \varrho \in \mathbb{R},$$

we obtain

$$\hat{A} = b_l x^{-\varkappa} (x^{-v(\lambda_l)} \partial - \Lambda_l) \dots (x^{-v(\lambda_{l(0)+1})} \partial - \Lambda_{l(0)+1}) \times \\ \times (\partial - \lambda_{l(0)}) \dots (\partial - \lambda_1) , \tag{2.2.6}$$

where $\Lambda_i \in \mathfrak{A}$ and $\Lambda_i(0) = \gamma_i$ $(i=l(0)+1,\dots,l)$. By equating the coefficients of D^i $(i \ge l(0))$ in (2.2.5) and (2.2.6), it is not difficult to get the result. □

2.2.2. The computation of the complete characteristic factors from the coefficients is much more complicated, though a finite algorithm for this is given by the method of proof of Theorem 2.1.1. In some cases, however, the problem may be reduced to the computation of the roots of a polynomial with coefficients in $\hat{\mathfrak{A}}$. With the formal differential operator (2.1.2) we associate the polynomial

$$p(x, \xi) = \sum_{0 \le i \le l} a_i(x) \xi^i .$$

It is well-known (cf. Krasnoselski et al. [1, Th. 21.6]) that there exist formal Puiseux series $\xi_i(x)$ such that

$$p(x, \xi) = \hat{a}_l(x)(\xi - \xi_l(x)) \dots (\xi - \xi_1(x)) .$$

The functions ξ_i are called branches of roots of p. Let

$$r_i = \max(1, 1 - v(x \xi_i))$$

and, for $\widetilde{x \xi_i} \neq 0$, denote the lowest term of $x \xi_i$ by $\eta_i x^{1-r_i}$.

Lemma 2.2.2 (Kannai [2]). For $i=1,\dots,l$, the lowest terms of $\tilde{\lambda}_i$ and $\widetilde{x\xi_i}$ coincide. If the condition

$$\eta_i \neq \eta_j \quad \text{for all } i \neq j \text{ such that } r_i = r_j > 1 \tag{2.2.7}$$

is satisfied, then the complete characteristic factors of $\hat{A}$ are given by $\lambda_i = \widetilde{x\xi_i}$ $(i=1,\dots,l)$.

Proof. 1. Let $q(i)$ and b_i as in (2.2.4) and (2.2.5), respectively, and $r_1 \le r_2 \le \dots \le r_l$. By equating the coefficients of ξ^i on both sides of the equality

$$p(x,\xi) = \hat{a}_l(\xi - \xi_1)\dots(\xi - \xi_l) = \sum_{0\le i\le l} b_i x^{q(i)-\varkappa} \xi^i ,$$

one obtains that $r_i = 1$ for $i=1,\dots,l_0$ and $r_i = q_i$, $\eta_i = \gamma_i$ for $i > l_0$ with q_i and γ_i defined in Lemma 2.2.1. This proves the first assertion of the lemma.

2. Assume that (2.2.7) holds and ξ_1 is a branch of roots of p such that $r_1 > 1$. Consider the operator

$$B = e^{-\lambda}\hat{A}e^{\lambda} , \quad \lambda = \int \widetilde{x\xi_1}x^{-1}dx ,$$

and denote its coefficients by $\hat{b}_i$. (2.1.5) implies

$$B = \hat{a}_l x^{-l}(\partial - \lambda_1 + \widetilde{x\xi_1}) \dots (\partial - \lambda_l + \widetilde{x\xi_1}) .$$

By (2.2.7) and the first assertion of the lemma, we thus obtain

$$\begin{aligned} v(\hat{b}_1) &= v(\hat{a}_l)-l+1 + \sum_{2\le i\le l} v(\lambda_i - \widetilde{x\xi_1}) \le v(\hat{a}_l)-l+1 \\ &+ (l-1)r_1 = v(\hat{a}_l) + (l-1)(r_1-1) . \end{aligned} \tag{2.2.8}$$

Furthermore, B can be written

$$\begin{aligned} B &= \sum_{0\le i\le l} \hat{a}_i \sum_{0\le j\le i} \binom{i}{j} e^{-\lambda}(e^{\lambda})^{(i-j)}D^j \\ &= \sum_{0\le j\le l} \ \sum_{j\le i\le l} \binom{i}{j}\hat{a}_j \, S(i-j,x)D^j, \end{aligned}$$

where $S(0,x) = 1$ and

$$S(k+1,x) = DS(k,x) + S(k,x)\widetilde{x\xi_1}x^{-1}, \quad k \in \mathbb{N}_0 .$$

By induction on k, we get

$$S(k,x) = [\xi_1(x)]^k + R_k(x), \quad v(R_k) \ge (k-1)(r_1-1) .$$

Since ξ_1 is a branch of roots of p and $\hat{b}_o = \hat{a}_o S(0,x)+\ldots+\hat{a}_l S(l,x)$,

$$v(\hat{b}_o) \geq v(\hat{a}_l) + (l-1)(r_1-1) - 1 .$$

Combining this with inequality (2.2.8), we find that $v(\hat{b}_o) \geq v(\hat{b}_1)-1$. Hence $\chi(B) < 1$ which means that $\widetilde{x\xi_1}$ is a characteristic factor of $\hat{A}$ (cf. Remark 2.1.7). □

If condition (2.2.7) is violated, then the second assertion of Lemma 2.2.2 does not hold, in general. For $l \leq 3$, we have

<u>Lemma 2.2.3</u> (Helffer and Kannai [1]). Let

$$p_1(x,\xi) = p(x,\xi)/\hat{a}_l(x),\ p_2(x,\xi) = p_1(x,\xi) - 2^{-1}\partial^2 p_1(x,\xi)/\partial x\,\partial\xi ,$$

$$p_3(x,\xi) = p_2(x,\xi) + 12^{-1}\partial^4 p_1(x,\xi)/\partial x^2\,\partial\xi^2 .$$

Then, for l=2,3, the complete characteristic factors of $\hat{A}$ are given by $\chi_i = \widetilde{x\xi_i}$ (i=1,...,l), where ξ_i are the branches of roots of p_l.

<u>Proof.</u> 1. Let l=2. We may assume $\hat{a}_2=1$. Setting

$$\hat{A} = D^2+\hat{a}_1 D+\hat{a}_o = [(D-\xi_1)(D-\xi_2)+(D-\xi_2)(D-\xi_1)]/2, \qquad (2.2.9)$$

we obtain $\xi_1+\xi_2 = -\hat{a}_1$ and $\xi_1\xi_2 = \hat{a}_o - D\hat{a}_1/2$. Thus $\hat{A}$ has the representation (2.2.9) if ξ_1 and ξ_2 are the branches of roots of p_2. Define

$$A_1 = (D-\xi_1)(D-\xi_2),\ A_2 = (D-\xi_2)(D-\xi_1) .$$

Both operators A_1 and A_2 have the characteristic factors $\widetilde{x\xi_i}$ (i=1,2). Let $\widetilde{x\xi_1}$, for example, be a characteristic factor of multiplicity k. Then, by Remark 2.1.7, $\chi(B_j) = 2-k$ (j=1,2), where

$$B_j = e^{-\lambda}A_j e^{\lambda} , \quad \lambda = \int \widetilde{x\xi_1}x^{-1}dx .$$

Since the coefficients of D and D^2 in B_1 coincide with those in B_2, we also have $\chi(B) = 2-k$, where $B = e^{-\lambda}\hat{A}e^{\lambda}$. Hence $\hat{A}$ possesses the characteristic factors $\widetilde{x\xi_1}$ and $\widetilde{x\xi_2}$.

2. Let l=3 and $\hat{a}_3=1$. Putting

$$\hat{A} = D^3+\hat{a}_2 D^2+\hat{a}_1 D+\hat{a}_o = [(D-\xi_1)(D-\xi_2)(D-\xi_3)$$
$$+ (D-\xi_1)(D-\xi_3)(D-\xi_2)+\ldots+(D-\xi_3)(D-\xi_2)(D-\xi_1)]/6 , \qquad (2.2.10)$$

we obtain

$$\zeta_1+\zeta_2+\zeta_3 = -\hat{a}_2, \quad \zeta_1\zeta_2+\zeta_1\zeta_3+\zeta_2\zeta_3 = \hat{a}_1-D\hat{a}_2,$$

$$\zeta_1\zeta_2\zeta_3 = -\hat{a}_0 + D\hat{a}_1/2 - D\hat{a}_2/6 .$$

Therefore $\hat{A}$ has the representation (2.2.1o) if $\zeta_1, \zeta_2, \zeta_3$ are the branches of roots of p_3. Define

$$A_1 = (D-\zeta_1)(D-\zeta_2)(D-\zeta_3) , \; A_2 = (D-\zeta_1)(D-\zeta_3)(D-\zeta_2),\ldots,$$

$$A_6 = (D-\zeta_3)(D-\zeta_2)(D-\zeta_1) .$$

Then all operators A_j $(j=1,\ldots,6)$ have the same characteristic factors $\widetilde{x\zeta_i}$ $(i=1,2,3)$. Let $\widetilde{x\zeta_1}$, for example, be a characteristic factor of multiplicity k and define B_j and B as above. Then $\chi(B_j) = 3-k$ for all j. We have to show that $\chi(B) = 3-k$. This is clear for k=2,3, since the coefficients of D^3 and D^2 in B_1 coincide with those in $B_2,\ldots,B_6$. Let k=1. Then $\widetilde{x\zeta_1} \neq \widetilde{x\zeta_i}$, i=2,3. Setting $\varsigma_i = \zeta_i - \widetilde{x\zeta_1}x^{-1}$, we obtain

$$B = (B_1+\ldots+B_6)/6 = D^3+b_2D^2+b_1D+b_o, \; b_2 = -(\varsigma_1+\varsigma_2+\varsigma_3) ,$$

$$b_1 = \varsigma_1\varsigma_2+\varsigma_1\varsigma_3+\varsigma_2\varsigma_3-Db_2 , \; b_o = \varsigma_1\varsigma_2\varsigma_3+Db_1/2-D^2b_2/6.$$

Since $v(\varsigma_1)\geq 0$ and $v(\varsigma_i) < -1$ (i=2,3), we finally obtain

$$v(b_1) = v(\varsigma_2\varsigma_3) < v(\varsigma_2+\varsigma_3)-1 \leq v(b_2)-1, \; v(b_o)\geq v(b_1)-1$$

which means $\chi(B)=2$ Q.E.D. □

We do not know how Lemma 2.2.3 generalizes to formal differential operators of higher order.

2.3. Principal parts of differential operators

2.3.1. Let A be the differential operator defined in (2.0.1). Consider the corresponding formal differential operator

$$\hat{A} = \sum_{0\leq i\leq 1} \hat{a}_i D^i , \; \hat{a}_i = \sum_{j\geq 0} a_i^{(j)}(0)x^j/j! . \tag{2.3.1}$$

By Corollary 2.1.2, $\hat{A}$ has a factorization of the form (2.1.5) with formal Puiseux series $\lambda_i(x)$ (i=1,...,l). For a rational number N, let λ_i^N be the finite sum of terms of order $<$ N in λ_i. The functions $\tilde{\lambda}_i = \lambda_i^o$, which are unique by Lemma 2.1.6, will be called the characteristic factors of A (at the origin). Moreover, the numbers

$$\mu_i = (\lambda_i - \tilde{\lambda}_i)(0),\ i=1,\dots,l\ ,$$

are unique when the order of the characteristic factors is fixed (cf. Remark 2.1.9). With A we associate a principal part

$$A_0 = x^{q-1}A_l \dots A_1\ ,\ A_i = \partial - \lambda_i^N(x)\quad (i=1,\dots,l)\ , \tag{2.3.2}$$

where $N > 0$ will be chosen sufficiently large later on. Furthermore, with the help of A_0, we define a weighted Sobolev space on $(0,b)$ in the following way. For $i=0,\dots,l$, let

$$B_i = A_iA_{i-1} \dots A_1\ ,\ w(i) = r(1)+ \dots +r(i),\ B_0 = I,\ w(0) = 0\ ,$$

where $r(i)$ denotes the largest non-positive rational number r such that $\mathrm{Re}\,\tilde{\lambda}_i(x)/x^r$ is bounded as $x \to 0+$. For $1 \le p \le \infty$ and $\rho \in \mathbb{R}$, let

$$X_p^\rho(0,b) = X_p^\rho(A_0;(0,b)) = \{y \in L_p^\rho(0,b) : B_iy \in L_p^{\rho+w(i)}(0,b), i=1,\dots,l\}\ ,$$

where $L_p^\rho = x^\rho L_p$ is the weighted L_p space defined in 1.1.1. It is easy to check that X_p^ρ is a Banach space with the norm

$$\|y\|_{X_p^\rho(0,b)} = \sum_{0 \le i \le l} \|B_iy\|_{L_p^{\rho+w(i)}(0,b)}\ , \tag{2.3.3}$$

since any Cauchy sequence with respect to (2.3.3) is also a Cauchy sequence in $W_p^1(\varepsilon,b)$ for all $\varepsilon \in (0,b)$. For brevity, we further set

$$Y_p^\rho(0,b) = L_p^{\rho+q-1+w(l)}(0,b)\ .$$

Then we obviously have $A_0 \in L(X_p^\rho(0,b),Y_p^\rho(0,b))$. Finally, let ξ be the number of characteristic factors of A satisfying $\mathrm{Re}\,\tilde{\lambda}_i(x) \to +\infty$ as $x \to 0+$ and $\xi(p,\rho)$ the number of indices i such that $r(i) = 0$ and $\mathrm{Re}\,\mu_i > -1/p+\rho+w(i-1)$, where $1/p=0$ for $p=\infty$. The following theorem is basic for the investigation of the degenerate operator (2.o.1).

<u>Theorem 2.3.1.</u> Let A_0 be a principal part of (2.o.1) of the form (2.3.2). Under the assumption

$$\mathrm{Re}\,\mu_i \ne -1/p+\rho+w(i-1) \text{ whenever } r(i)=0\ (1 \le i \le l), \tag{2.3.4}$$

there exists $b_0 > 0$ such that, for $b \in (0,b_0]$, $A_0 \in L(X_p^\rho(0,b),\ Y_p^\rho(0,b))$ is right invertible with kernel index $\xi + \xi(p,\rho)$. Moreover, there exist right inverses $A^{-1}_{o,b}$ of A_o on $(0,b)$ such that

$$\|A^{-1}_{o,b}\|_{L(Y^{\varrho}_p(0,b),X^{\varrho}_p(0,b))} \le c\ ,\ b\in (0,b_o]\ . \tag{2.3.5}$$

<u>2.3.2.</u> First, we shall verify Theorem 2.3.1 for a first order operator

$$B = \partial - \lambda(x),\ \lambda = \sum_{M\le j\le N} c_j x^{j/m}\ ,\ c_j\in\mathbb{C}\ ,\ m\in\mathbb{N}\ ,$$
$$N,M\in\mathbb{Z}\ ,\ M\le 0\le N. \tag{2.3.6}$$

Let $w = \bar{j}/m$ be the largest non-positive rational number r such that $\operatorname{Re}\lambda(x)/x^r$ is bounded as $x\to 0+$. Then

$$\operatorname{Re} c_j = 0\ (j<\bar{j})\quad\text{and, for } w<0,\quad \operatorname{Re} c_{\bar{j}} \ne 0\ . \tag{2.3.7}$$

Condition (2.3.4) for B reads

$$\operatorname{Re} c_o \ne -1/p + \varrho \quad \text{when } w = 0\ . \tag{2.3.8}$$

We consider the cases

(i) $\operatorname{Re} c_{\bar{j}} < 0$ for $w<0$, $\operatorname{Re} c_o < -1/p+\varrho$ for $w=0$;

(ii) $\operatorname{Re} c_{\bar{j}} > 0$ for $w<0$, $\operatorname{Re} c_o > -1/p+\varrho$ for $w=0$.

In case (i) we set

$$(B^{-1}_b f)(x) = e^{h(x)}\int_0^x e^{-h(z)}z^{-1}f(z)dz\ , \tag{2.3.9}$$

and in case (ii)

$$(B^{-1}_b f)(x) = e^{h(x)}\int_b^x e^{-h(z)}z^{-1}f(z)dz\ , \tag{2.3.10}$$

where

$$h(x) = \sum_{M\le j\le N,\, j\ne 0} (m/j)c_j x^{j/m} + c_o \ln x\ .$$

Note that the function under the integral sign in (2.3.9) is integrable on $(0,x)$. Let $\bar{X}^{\varrho}_p(0,b) = X^{\varrho}_p(0,b)$ in case (i) and $\bar{X}^{\varrho}_p(0,b) =$ $= \{\, y\in X^{\varrho}_p(0,b)\ :\ y(b) = 0\,\}$ in case (ii), where as above

$$X^{\varrho}_p(0,b) = X^{\varrho}_p(B;(0,b)) = \{\, y\in L^{\varrho}_p(0,b)\ :\ By\in L^{\varrho+w}_p(0,b)\}\ .$$

<u>Theorem 2.3.2.</u> Under hypothesis (2.3.8), there exists a number $b_o>0$ such that for $b\in(0,b_o]$, the equation

$$By = f,\ f\in L^{\varrho+w}_p(0,b) \tag{2.3.11}$$

has always the unique solution $y = B_b^{-1}f$ in $\bar{X}_p^{\varrho}(0,b)$ satisfying

$$c \|y\|_{L_p^{\varrho}(0,b)} \leq \|f\|_{L_p^{\varrho+w}(0,b)} \tag{2.3.12}$$

with c independent of f and b.

<u>Proof.</u> 1. For $\varrho \neq 0$, we may pass to the operator $B_\varrho = x^{-\varrho}Bx^{\varrho} = B+\varrho$. Applying Theorem 2.3.2 (for $\varrho = 0$) to B_ϱ, we obtain the conclusion for B. Thus we may assume $\varrho = 0$ in the sequel.

2. Let $\bar{W}_p^1(0,b)$ be the set of all functions y in $\mathring{W}_p^1(0,b)$ which vanish in some neighborhood of x=0 and satisfy y(b)=0 in case (ii). For all $y \in \bar{W}_p^1(0,b)$, $p \geq 2$ and sufficiently small b, we now verify the estimate

$$\|x^{-w}By\|_{L_p(0,b)} \geq c\|y\|_{L_p(0,b)} \tag{2.3.13}$$

(in the sequel $c, c_1, \ldots$ denote positive constants independent of y and b). Let first p=2. Setting $q = 1-w = 1-\bar{j}/m$, we obtain

$$\int_0^b x^q\bar{y}Dydx = -\int_0^b x^q y\overline{Dy}dx - q\int_0^b x^q|y|^2dx + b^q|y(b)|^2,$$

hence

$$\mathrm{Re}\int_0^b x^q\bar{y}Dydx = 2^{-1}\{b^q|y(b)|^2 - q\int_0^b x^q|y|^2dx\},$$

$$\mathrm{Re}\int_0^b x^{-w}\bar{y}Bydx = -\mathrm{Re}\, c_{\bar{j}}\|y\|^2_{L_2(0,b)} - 2^{-1}q\|x^{q/2}y\|^2_{L_2(0,b)}$$
$$+ 2^{-1}b^q|y(b)|^2 + \sum_{\bar{j}+1 \leq j \leq N}\int_0^b \mathrm{Re}\, c_j x^{j/m-w}|y|^2dx \tag{2.3.14}$$

in view of (2.3.7). In case (i) it follows from (2.3.14) and $\mathrm{Re}\, c_{\bar{j}} < 0$ for $q > 1$ (resp. $\mathrm{Re}\, c_0 + 1/2 < 0$ for q=1) that

$$|\mathrm{Re}\int_0^b x^{-w}\bar{y}Bydx| \geq c_1\|y\|^2_{L_2(0,b)}, \quad y \in \bar{W}_2^1(0,b), \tag{2.3.15}$$

when b is sufficiently small. In case (ii) from (2.3.14) we deduce (2.3.15) again. Combining (2.3.15) with the Cauchy-Schwarz inequality, we obtain (2.3.13) for p=2.

Let now $p \in (2,\infty)$. For $y \in \bar{W}_p^1(0,b)$, we have $|y|^{p/2} \in \bar{W}_2^1(0,b)$ since

$$(|y|^{p/2})'(x) = \begin{cases} 2^{-1}p|y|^{p/2-2}\,\mathrm{Re}\, y'\bar{y} & \text{if } y(x)\neq 0, \\ 0 & \text{if } y(x)=0. \end{cases}$$

Furthermore

$$\left|\operatorname{Re} 2^{-1}p \int_0^b |y|^{p-2}x^{-w}\bar{y}By\,dx\right| = \left|\operatorname{Re} \int_0^b x^{-w}|y|^{p/2}B_1|y|^{p/2}dx\right| ,$$

where $B_1 = \partial - 2^{-1}p\,\lambda(x)$. Since B_1 satisfies assumption (2.3.8) for $\varrho=0$ and $p=2$, by (2.3.15) the inequality

$$\left|\operatorname{Re} \int_0^b x^{-w}|y|^{p/2}B_1|y|^{p/2}dx\right| \geq c_2 \|y^{p/2}\|^2_{L_2(0,b)} = c_2\|y\|^p_{L_p(0,b)}$$

holds when b is sufficiently small. Together with the estimate

$$\left|\operatorname{Re} \int_0^b |y|^{p-2}x^{-w}\bar{y}By\,dx\right| \leq \|x^{-w}By\|_{L_p(0,b)}\|y^{p-1}\|_{L_{\bar{p}}(0,b)}$$

$$= \|x^{-w}By\|_{L_p(0,b)}\|y\|^{p-1}_{L_p(0,b)} \quad , \quad 1/p+1/\bar{p}=1 ,$$

we obtain (2.3.13).

Let $p=\infty$. Then condition (2.3.8) is satisfied for all sufficiently large p and an inspection of the above proof shows that (2.3.13) is valid with c independent of $y \in \bar{W}^1_\infty(0,b)$ and $p \geq p_0$, $b \leq b_0$ for some $p_0 \geq 2$ and $b_0 > 0$. Using the relation $\|y\|_{L_p} \to \|y\|_{L_\infty}$ as $p \to \infty$, we obtain (2.3.13) for $p=\infty$.

3. We now complete the proof of the theorem in the case $p \geq 2$. First, we observe that $\bar{W}^1_p(0,b)$ is dense in $\bar{X}^0_p(0,b)$. Indeed, setting $y_n(x) = \chi(nx)y(x)$ for $n \in \mathbb{N}$ and $y \in \bar{X}^0_p(0,b)$, where χ is a smooth function on $\mathbb{R}$ such that $0 \leq \chi \leq 1$, $\chi=0$ for $|x| \leq 1/2$ and $\chi = 1$ for $|x| \geq 1$, we obtain $y_n \in \bar{W}^1_p(0,b)$ $(n \geq n_0)$ and $y_n \to y$, $x^{1-w}yD\chi(nx) \to 0$ and $\chi(nx)x^{-w}By \to x^{-w}By$ $(n \to \infty)$ in $L_p(0,b)$, hence $y_n \to y$ in $X^0_p(0,b)$. Thus inequality (2.3.13) holds for all $y \in \bar{X}^0_p(0,b)$ and sufficiently small b. Furthermore, for any $f \in C^\infty_0(0,b)$, it is easy to check that $B_b^{-1}f$ is a solution of equation (2.3.11) in $\bar{X}^0_p(0,b)$. Since $C^\infty_0(0,b)$ is dense in $L^w_p(0,b)$, it follows from inequality (2.3.13) that (2.3.11) has always a unique solution in $\bar{X}^0_p(0,b)$ satisfying (2.3.12) when b is sufficiently small. It remains to show that the solution takes the form $B_b^{-1}f$ for any $f \in L^w_p$. Let $\{f_n\} \subset C^\infty_0(0,b)$ be a sequence such that $f_n \to f$ $(n \to \infty)$ in $L^w_p(0,b)$. Then one can deduce from (2.3.9) and (2.3.1o) that $(B_b^{-1}f_n)(x) \to (B_b^{-1}f)(x)$ as $n \to \infty$ for all $x > 0$ which proves the assertion.

4. Finally, we consider the case $p \in [1,2)$. Let $\bar{B} = -\partial - \bar{\lambda} - 1$ be the formal adjoint of B. $\bar{B}$ satisfies assumption (2.3.8) with $\varrho = -w$ and p replaced by $\bar{p} = p/(p-1)$. Let $\bar{B}_b^{-1}$ be the corresponding right inverses (2.3.9) or (2.3.10) of $\bar{B}$. Then we obtain inequality (2.3.13) for all $y \in \bar{W}_p^1(0,b)$ and sufficiently small b:

$$\| x^{-w} By \|_{L_p} \geq c \| x^{-w} By \|_{L_p} \sup \| x^w \bar{B}_b^{-1} v \|_{L_{\bar{p}}} \| v \|_{L_{\bar{p}}}^{-1}$$

$$\geq c \sup | (x^{-w} By, x^w \bar{B}_b^{-1} v) | \, \| v \|_{L_{\bar{p}}}^{-1} = c \sup |(y,v)| \, \| v \|_{L_{\bar{p}}}^{-1}$$

$$= c \| y \|_{L_p} ,$$

where the supremum is taken over all $v \in L_{\bar{p}}$ such that $v \neq 0$ and $(.,.)$ denotes the scalar product in $L_2(0,b)$. The rest of the proof is the same as for $p \geq 2$. □

Remark 2.3.3. Under hypothesis (2.3.8) for $p = \infty$, Theorem 2.3.2 remains valid if the L_∞^ϱ spaces are replaced by $\mathring{C}^\varrho = \{ x^\varrho y \; : \; y \in C^0, y(0)=0 \}$. It is sufficient to verify that, for $f \in \mathring{C}^w$, the solution $y = B_b^{-1} f$ of (2.3.11) belongs to $\mathring{C}^0$. Indeed, in case (i) we obtain by l'Hospital's rule

$$\lim_{x \to 0+} |e^{h(x)}| \int_0^x | e^{-h(z)} z^{-1} f(z) | dz$$

$$= \lim_{x \to 0+} \int_0^x | e^{-\mathrm{Re}\, h(z)} z^{-1} f(z) | dz / e^{-\mathrm{Re}\, h(x)}$$

$$= \lim_{x \to 0+} \{ -x^{-1} f(x) / x^{-1} \mathrm{Re}\, (\varkappa) \} = 0 ,$$

hence $(B_b^{-1} f)(x) \to 0$ as $x \to 0+$. In case (ii), the proof is analogous.

2.3.3. Proof of Theorem 2.3.1. Consider the equation

$$A_0 y = x^{q-1} A_l \ldots A_1 y = f, \; f \in Y_p^\varrho(0,b) . \qquad (2.3.16)$$

Since each operator A_i ($i=1,\ldots,l$) satisfies condition (2.3.8) with ϱ replaced by $\varrho + w(i-1)$, it follows from Theorem 2.3.2 that, for all sufficiently small b, (2.3.16) has the unique solution

$$y = A_{0,b}^{-1} f := A_{1,b}^{-1} \ldots A_{l,b}^{-1} x^{1-q} f$$

in

$$\bar{X}_p^\varrho(0,b) = \{ y \in X_p^\varrho(0,b) : (B_{i-1}y)(b) = 0, \ i \in M \} \quad ,$$

where $A_{i,b}^{-1}$ are the corresponding right inverses of A_i and M is the set of all indices i satisfying condition (ii). Moreover, applying the corresponding estimates (2.3.12) for A_i, we obtain

$$\| A_{o,b}^{-1} f \|_{X_p^\varrho(0,b)} = \sum_{0 \le i \le l} \| B_i A_{o,b}^{-1} f \|_{L_p^{\varrho+w(i)}(0,b)}$$

$$= \sum_{0 \le i \le l} \| A_{i+1,b}^{-1} \cdots A_{l,b}^{-1} x^{l-q} f \|_{L_p^{\varrho+w(i)}(0,b)} \le c \| f \|_{Y_p^\varrho(0,b)}$$

with c independent of f and b. Thus $A_{o,b}^{-1}$ is the desired right inverse of A_o. By virtue of the relation $\dim X_p^\varrho / \bar{X}_p^\varrho = \xi + \xi(p,\varrho)$, we finally conclude the result on the nullity of A_o. □

2.4. The index in weighted L_p spaces and C^∞

2.4.1. We first study the operator A defined in (2.o.1) in weighted L_p spaces. Preserving the notation of the preceding sections, we assume that the characteristic factors $\tilde{\lambda}_i$ of A are labelled in such a way that (2.2.1) holds. Let $q(i)$ be the numbers defined by (2.2.4). We now choose a principal part A_o of the form (2.3.2), where N is so large that

$$x^{l-q-w(l)}(\bar{a}_l^{-1}A - A_o) = e_o + \dots + e_{l-1}D^{l-1}, e_i \in C^o[0,b],$$
$$x^{-q(i)} e_i(x) \to 0 \ (x \to 0), \ i=0,\dots,l-1 \ . \qquad (2.4.1)$$

Theorem 2.4.1. Under hypotheses (2.3.4) and (2.4.1), the operator $A \in L(X_p^\varrho(A_o;(0,b)), Y_p^\varrho(0,b))$ is right invertible with kernel index $\xi + \xi(p,\varrho)$.

Proof. It suffices to show that

$$\| \bar{a}_l^{-1} A - A_o \|_{L(X_p^\varrho(0,b), Y_p^\varrho(0,b))} \to 0 \quad \text{as } b \to 0 \ . \qquad (2.4.2)$$

Indeed, as in the proof of Theorem 1.3.1(i) we then obtain the result from Theorem 2.3.1.

Recall that $q(i) = q_o + \dots + q_i$, where $q_o = 0$ and $q_i = \max(1, 1 - v(\tilde{\lambda}_i))$. Since $w(i) \ge i - q(i)$ and $q_i \ge q_{i-1}$, one successively obtains

$$\| B_i\, y \|_{L_p^{\varrho+w(i)}(0,b)} \geq \| x^{q(i)-i} B_i y \|_{L_p^{\varrho}(0,b)} \geq \| x^{q(i)} D^i y \|_{L_p^{\varrho}(0,b)}$$

$$- c \sum_{0 \leq j < i} \| x^{q(j)} D^j y \|_{L_p^{\varrho}(0,b)} , \quad i=0,\dots,l ,$$

for all $y \in X_p^{\varrho}(0,b)$ and $b \leq 1$. This gives the inequality

$$\sum_{0 \leq i \leq l} \| x^{q(i)} D^i y \|_{L_p^{\varrho}(0,b)} \leq c_1 \| y \|_{X_p^{\varrho}(0,b)} \tag{2.4.3}$$

with c_1 independent of y and b. Now (2.4.2) is an easy consequence of (2.4.1) and (2.4.3). □

Corollary 2.4.2. For $\varrho > q-1$ and any right-hand side $f \in L_p^{\varrho}(0,b)$, the differential equation $Ay=f$ has a solution in $L_p(0,b)$.

Proof. We may choose $\bar{\varrho} \in [0, \varrho-q+1]$ such that condition (2.3.4) is satisfied with ϱ replaced by $\bar{\varrho}$. Since $q-1+w(1) \leq q-1$, we have $L_p^{\varrho} \subset Y_p^{\bar{\varrho}}$ and the assertion follows from Theorem 2.4.1. □

For $\varrho = q-1$, Corollary 2.4.2 does not hold, in general (cf. Theorems 1.3.1(ii) and 1.7.1).

Remark 2.4.3. The spaces $\overset{\circ}{X}{}_p^{\varrho}(0,b) = X_p^{\varrho}(A_o;(0,b))$ depend on the choice of the principal part A_o, in general. However, these spaces coincide when the number N in (2.3.2) is sufficiently large. This is easily seen by virtue of estimate (2.4.3).

Example 2.4.4. If A is of first order, then we have

$$X_p^{\varrho}(0,b) = \{ y \in L_p^{\varrho}(0,b) : Ay \in L_p^{\varrho+\bar{j}}(0,b) \}$$

$$= \{ y \in L_p^{\varrho}(0,b) : A_o y \in L_p^{\varrho+\bar{j}}(0,b) \} ,$$

where

$$A_o = x^q D + \sum_{0 \leq j \leq \bar{j}} b^{(j)}(0) x^j / j! \, , \quad b = a_o / \bar{a}_1 ,$$

$$\bar{j} = \min(q-1, v(\mathrm{Re}\, [\, b(0) + \dots + b^{(q-1)}(0) x^{q-1}])) .$$

By Theorem 2.4.1, $A \in L(X_p^{\varrho}(0,b), L_p^{\varrho+\bar{j}}(0,b))$ is normally solvable if

$$\mathrm{Re}\; b^{(q-1)}(0)/(q-1)! \neq -1/p+\varrho \quad \text{when } \bar{j}=q-1 . \tag{2.4.4}$$

Using the method of proof of Theorem 1.3.1(ii), one can show that

$A \in L(X_p^{\varrho}, L_p^{\varrho+\bar{j}})$ is not normally solvable if (2.4.4) is violated. Moreover, the operator $A = \bar{a}_1 x^q D + a_0$ with domain of definition

$$D(A) = \{ y \in L_p^{\varrho}(0,b) : Ay \in L_p^{\bar{\varrho}}(0,b) \}$$

is not normally solvable as a map of L_p^{ϱ} into $L_p^{\bar{\varrho}}$ when $\bar{\varrho} < \bar{j} + \varrho$. Consider the operators

$$A(\varepsilon) = A + \varepsilon x^{\bar{\varrho}} : D(A) \to L_p^{\bar{\varrho}}(0,b), \quad \varepsilon \in \mathbb{R}, \ \bar{\varrho} < \bar{j} + \varrho .$$

Using the arguments from the proof of Theorem 2.3.2, one can verify that $A(\varepsilon)$ has index 0 for $\varepsilon > 0$ and index 1 for $\varepsilon < 0$. Then, as in the proof of Theorem 1.3.1(ii), we see that $A = A(0)$ cannot be normally solvable.

Thus we observe that the choice of the weighted spaces X_p^{ϱ}, Y_p^{ϱ} in Theorem 2.4.1 is optimal at least for first order operators.

<u>2.4.2.</u> In order to investigate the solvability properties of

$$Ay = f, \ f \in C^{\infty}[a,b], \quad 0 \in [a,b] , \tag{2.4.5}$$

we consider the equation

$$\hat{A}\,\hat{y} = \hat{f}, \quad \hat{f} = \sum_{j \geq 0} f^{(j)}(0) x^j / j! \tag{2.4.6}$$

in the space $\mathbb{C}[[x]]$ of formal power series in the indeterminate x, where $\hat{A}$ is the corresponding formal differential operator (2.3.1). Define $\varkappa = \varkappa(A) = \varkappa(\hat{A})$ as in (2.1.4) and let

$$k_0 = \min \{ k \in \mathbb{N}_0 : P(j) \neq 0 , \ j = k, k+1, \ldots \} ,$$

where P denotes the characteristic polynomial (2.1.12). For $l_0 = 0$, we have $P(\mu) = (x^{\varkappa} a_0)(0) \neq 0$, hence $k_0 = 0$. Note thate $k_0 \geq \varkappa$ since, for $\varkappa > 0$, (2.1.12) has the roots $0, 1, \ldots, \varkappa - 1$.

With (2.4.5) we further associate the finite linear system

$$(\hat{A}\hat{y})^{(j)}(0) = f^{(j)}(0) , \ j = 0, \ldots, k_0 - \varkappa - 1, \ \hat{y} \in \mathbb{C}[[x]], \tag{2.4.7}$$

in the unknowns $y_j = \hat{y}^{(j)}(0)$, $j = 0, \ldots, k_0 - 1$.

<u>Lemma 2.4.5.</u> Equation (2.4.6) is solvable in $\mathbb{C}[[x]]$ if and only if system (2.4.7) is solvable. Moreover, for each solution $(y_j)_0^{k_0-1}$ of (2.4.7), there exists a unique formal power series $\hat{y}$ satisfying (2.4.6)

and $\hat{y}^{(j)}(0) = y_j$ $(j=0,\dots,k_0-1)$. For $k_0=0$, (2.4.7) means $f^{(j)}(0) = 0$ $(j=0,\dots,-\varkappa-1)$ and each solution of (2.4.6) is unique. If $k_0 = \varkappa$, then (2.4.7) is void and, for any right-hand side and any vector $(y_j)_0^{\varkappa-1}$, there exists a unique $\hat{y} \in \mathbb{C}[[x]]$ such that (2.4.6) and $\hat{y}^{(j)}(0) = y_j$ $(j=0,\dots,\varkappa-1)$ hold.

Proof. (2.4.6) is obviously equivalent to the infinite linear system $(\hat{A}\hat{y})^{(j)}(0) = f^{(j)}(0)$ $(j \in \mathbb{N}_0)$ in the unknowns $\hat{y}(0), \hat{y}'(0), \dots$. For all $k \geq k_0$, consider the k-by-k matrices $\mathcal{A}_k$ defined by

$$\mathcal{A}_k Y = ((x^{\varkappa}\hat{A}\hat{y})^{(j)}(0))_0^{k-1}, \quad Y = (\hat{y}^{(j)}(0))_0^{k-1} .$$

We observe that $\mathcal{A}_k$ is in lower triangular form and the diagonal element in the jth row is given by $P(j-1)$. In view of $P(k) \neq 0$ for all $k \geq k_0$, it is easily seen that the lemma holds when $\varkappa=0$. Since the matrices of the systems (2.4.7) with k_0 replaced by k take the form

$$\begin{pmatrix} \mathbf{0} \\ \mathcal{D}\mathcal{A}_k \end{pmatrix} \text{ when } \varkappa < 0, \quad (\mathcal{B} \;\; \mathcal{A}_k\bar{\mathcal{D}}) \text{ when } \varkappa > 0$$

with the $(-\varkappa)$-by-k zero matrices $\mathbf{0}$, some k-by-$\varkappa$ matrices $\mathcal{B}$ and invertible diagonal matrices $\mathcal{D}$ and $\bar{\mathcal{D}}$, we obtain the lemma in the general case. □

Let n_0 be the kernel index of the matrix of system (2.4.7) considered as a linear map of $\mathbb{C}^{k_0}$ into $\mathbb{C}^{k_0-\varkappa}$. Let further ξ (resp. ξ') be the number of the characteristic factors $\tilde{\lambda}_i$ of A satisfying $\operatorname{Re} \tilde{\lambda}_i(x) \to +\infty$ as $x \to 0+$ (resp. $x \to 0-$). For b=0 and a=0, set $\xi=0$ and $\xi'=0$, respectively. Now we can state

Theorem 2.4.6. $A \in L(C^{\infty}[a,b])$ is a Fredholm operator with index $\xi + \xi' + \varkappa$ and nullity $\xi + \xi' + n_0$. Moreover, (2.4.5) is solvable in $C^{\infty}[a,b]$ if and only if (2.4.7) is solvable.

Proof. 1. Let $C^{\infty,0}[a,b]$ be the closed subspace $\{y \in C^{\infty}[a,b] : y^{(j)}(0) = 0,\ j \in \mathbb{N}_0\}$ of flat functions of $C^{\infty}[a,b]$. We first consider the operator (2.o.1) in $C^{\infty,0}[0,b]$, $b > 0$.

Let us introduce the spaces

$$X^{\infty} = \bigcap_{\rho \geq 0} X_2^{\rho}(0,b), \quad Y^{\infty} = \bigcap_{\rho \geq 0} Y_2^{\rho}(0,b) = \bigcap_{\rho \geq 0} L_2^{\rho}(0,b)$$

endowed with the topology of the projective limit of the Banach spaces X_2^ϱ, Y_2^ϱ defined above (cf. Robertson and Robertson [1]). For all sufficiently large ϱ, assumption (2.3.8) (with p=2) is satisfied and $\xi(2,\varrho) = 0$. By Theorem 2.4.1, $A \in L(X_2^\varrho(0,b),Y_2^\varrho(0,b))$ is a surjective operator with nullity ξ whenever ϱ is large enough. Hence $A \in L(X^\infty,Y^\infty)$ is also surjective with kernel index ξ. Moreover, it follows from $y \in X^\infty$ and $Ay \in C^{\infty,0}[0,b]$ that $y \in C^{\infty,0}[0,b]$. Indeed, $y \in X^\infty$ implies $D^i y \in Y^\infty$ $(i=0,\dots,l)$, and by differentiating Ay for $x \neq 0$, we further obtain $D^{l+1}y \in Y^\infty$. By iteration, $D^i y \in Y^\infty$ for all $i > l+1$ which proves the assertion. Consequently, A is a surjective operator with nullity ξ in $C^{\infty,0}[0,b]$.

2. In an analogous manner one obtains that A is surjective in $C^{\infty,0}[a,0]$ $(a<0)$ with nullity ξ'. Thus $A \in L(C^{\infty,0}[a,b])$ is a surjective operator with nullity $\xi + \xi'$.

3. We now show that, for any $\hat{y} \in \mathbb{C}[[x]]$ satisfying (2.4.6), there exists a solution $y \in C^\infty[a,b]$ of (2.4.5) satisfying $y^{(j)}(0) = \hat{y}^{(j)}(0)$ for all $j \in \mathbb{N}_0$. By Borel's theorem (cf. Narasimhan [1]), there exists a function $y_1 \in C^\infty[a,b]$ such that $y_1^{(j)}(0) = \hat{y}^{(j)}(0)$, $j \in \mathbb{N}_0$. Hence $g = f - Ay_1 \in C^{\infty,0}[a,b]$. Choosing a solution $y_2 \in C^{\infty,0}[a,b]$ of the equation $Ay = g$, we conclude that $y_1 + y_2$ is the desired solution of (2.4.5). Together with Lemma 2.4.5, this yields that (2.4.5) is solvable in $C^\infty[a,b]$ if and only if (2.4.7) is solvable and, for each solution $(y_j)_0^{k_0-1}$ of (2.4.7), there exists a solution $y \in C^\infty[a,b]$ of (2.4.5) satisfying $y^{(j)}(0) = y_j$ $(j=0,\dots,k_0-1)$ which is unique modulo $C^{\infty,0}[a,b]$. Hence $\dim C^\infty[a,b]/A(C^\infty[a,b]) = n_0 - \varkappa$, and by virtue of the result in 2., $\dim \ker A = \xi + \xi' + n_0$ in $C^\infty[a,b]$ which completes the proof of the theorem. □

<u>Corollary 2.4.7.</u> If $\varkappa \geq 0$ and $P(j) \neq 0$ $(j=0,\dots,k_0-1)$, then A is surjective in $C^\infty[a,b]$. For $\varkappa < 0$, A cannot be surjective since the solvability conditions $f^{(j)}(0) = 0$ $(j=0,\dots,-\varkappa-1)$ occur.

Finally, as a simple example, we consider the second order operator $A = x^2D^2 + \alpha D + \beta$, $\alpha, \beta \in \mathbb{C}$. For $\alpha = 0$, we have $l_0 = 2$, $\varkappa = 0$ and both characteristic factors vanish identically. If $\alpha \neq 0$, then $l_0 = \varkappa = 1$ and A has exactly one non-vanishing characteristic factor which is $-\alpha x^{-1}$. This is easily seen by equating the coefficients of D on both sides of the equation $A = \hat{A} = (\partial - \lambda_2)(\partial - \lambda_1)$, where $\lambda_1, \lambda_2 \in \hat{\mathcal{O}}$. Thus, by Theorem 2.4.6, $A \in L(C^\infty[0,b])$ is a Fredholm operator with

$$\operatorname{ind} A = 0 \quad \text{when} \quad \alpha = 0, \ \operatorname{ind} A = \begin{cases} 2 \text{ when } \operatorname{Re} \alpha < 0, \\ 1 \text{ when } \operatorname{Re} \alpha \geq 0 \text{ and } \alpha \neq 0 . \end{cases}$$

We see that the index of the operator (2.o.1) in C^∞ depends on the lower order terms in a rather complicated way, in general.

2.5. Hypoellipticity and basis of solutions

2.5.1. We first compute the index of the operator A defined by (2.o.1) in the space $\mathcal{D}'(a,b)$ of all distributions in (a,b), where $a < 0 < b$. Preserving the notation of the preceding sections, by ζ (resp. ζ') we denote the number of characteristic factors $\tilde{\lambda}_i$ of A satisfying $\operatorname{Re} \tilde{\lambda}_i(x) \to -\infty$ as $x \to 0+$ (resp. $x \to 0-$).

Theorem 2.5.1. $A \in L(\mathcal{D}'(a,b))$ is a surjective Fredholm operator with kernel index $2l - \varkappa - \zeta - \zeta'$.

Proof. It is easy to check that the transpose ${}^tA$ of A has the characteristic factors $-\tilde{\lambda}_i$, $i=1,\dots,l$. By Theorem 2.4.6, ${}^tA \in L(C^\infty[a,b])$ is a Fredholm operator with index $\zeta + \zeta' + \varkappa$. We introduce the subspaces

$$Z^n[a,b] = \{y \in C^\infty[a,b] : y^{(j)}(a) = y^{(j)}(b) = 0, \ j=0,\dots,n\},$$

$$Z^\infty[a,b] = \bigcap_{n \geq 0} Z^n[a,b]$$

of $C^\infty[a,b]$. Since

$$\dim C^\infty[a,b]/Z^n[a,b] = 2(n+1),$$

${}^tA \in L(Z^{n+1}[a,b], Z^n[a,b])$ is a Fredholm operator with index $\zeta + \zeta' + \varkappa - 2l$ for any $n \in \mathbb{N}_0$ which is clearly injective in view of the classical uniqueness theorem for ordinary differential equations. Therefore ${}^tA \in L(Z^\infty[a,b])$ is an injective Fredholm operator with index $\zeta + \zeta' + \varkappa - 2l$. Now we observe that $C_0^\infty(a,b)$ is the inductive limit of the spaces $Z^\infty[a+1/n, b-1/n], n \geq n_0$ (cf. Robertson and Robertson [1]). Thus ${}^tA$ is also an injective operator with index $\zeta + \zeta' + \varkappa - 2l$ in $C_0^\infty(a,b)$.

Since $\mathcal{D}'$ is the dual space of C_0^∞, $A \in L(\mathcal{D}'(a,b))$ is a surjective operator with index $-(\zeta+\zeta'+\varkappa-2l)$. □

The differential operator (2.o.1) is called hypoelliptic at the origin if for every $y \in \mathcal{D}'(a,b)$, $Ay \in C^\infty[a,b]$ implies $y \in C^\infty[a,b]$. Let l_0 be the number (2.1.3) and $\tilde{\lambda}_i$ $(i=l_0+1,\dots,l)$ the non-vanishing characteristic factors of A.

<u>Theorem 2.5.2.</u> A is hypoelliptic at the origin if and only if $a_{l_0}(0)\neq 0$ and $|\mathrm{Re}\,\tilde{\lambda}_i(x)| \to \infty$ as $x \to 0$ for $i=l_0+1,\dots,l$.

<u>Proof.</u> First, we verify that A is hypoelliptic at the origin if and only if the indices of A in $C^\infty[a,b]$ and $\mathcal{D}'(a,b)$ coincide. Let A be hypoelliptic. Then the kernels of A in C^∞ and $\mathcal{D}'$ coincide and $A(C^\infty) = C^\infty$ since $A(\mathcal{D}') = \mathcal{D}'$. Conversely, if A has the same index in C^∞ and $\mathcal{D}'$, then the kernels of A in C^∞ and $\mathcal{D}'$ and the linear spaces $A(C^\infty)$ and $A(\mathcal{D}')\cap C^\infty$ must coincide since C^∞ is dense in $\mathcal{D}'$ (see the proof of Theorem 1.5.1). Hence A is hypoelliptic.

Furthermore, by Theorems 2.4.6 and 2.5.1, the indices of A in C^∞ and $\mathcal{D}'$ coincide if and only if the equality

$$2l = \xi+\zeta+\xi'+\zeta'+2\varkappa \tag{2.5.1}$$

holds. Since by definition $\varkappa \le l_0$, $\xi+\zeta \le l-l_0$ and $\xi'+\zeta' \le l-l_0$, (2.5.1) is equivalent to the three equalities

$$\varkappa = l_0\ ,\quad \xi+\zeta = l-l_0,\quad \xi'+\zeta' = l-l_0\ . \tag{2.5.2}$$

Finally, (2.5.2) is equivalent to $a_{l_0}(0) \neq 0$, $|\mathrm{Re}\,\tilde{\lambda}_i(x)| \to \infty$ as $x \to 0+$ and $|\mathrm{Re}\,\tilde{\lambda}_i(x)| \to \infty$ as $x \to 0-$ $(i=l_0+1,\dots,l)$. □

<u>2.5.2.</u> As a further application of the results in 2.1 and 2.4 we study the asymptotic behaviour of the basis of solutions to the homogeneous differential equation

$$Ay = \sum_{0\le i\le l} a_i(x)D^i y = 0\ ,\quad 0 < x \le b\ . \tag{2.5.3}$$

Let $\tilde{\lambda}_i \in \hat{\mathfrak{N}}_m$ $(i=1,\dots,\bar{l}\le l)$ be the different characteristic factors of A with multiplicities k_i. As in 2.1.3, set $\hat{\lambda}_i = \int \tilde{\lambda}_i x^{-1}dx$. Furthermore, let μ_{ij} $(j=1,\dots,\bar{k}_i \le k_i)$ be the different roots of the characteristic equation of the operator $\hat{A}_i = e^{-\hat{\lambda}_i}\hat{A}e^{\hat{\lambda}_i}$ with multiplicities r_{ij} (cf. (2.1.12) and Remark 2.1.9), where $\hat{A}$ is defined by (2.3.1).

<u>Theorem 2.5.3.</u> (2.5.3) has a basis of solutions of the form

$$y_{ijk}(x) = e^{\hat{\lambda}_i(x)} x^{\mu_{ij}}[(\ln x)^k + \sum_{0\leq n\leq s} \varphi_n(x^{1/m})(\ln x)^n], \tag{2.5.4}$$

$$k=0,\dots,r_{ij}-1,\ j=1,\dots,\bar{k}_i,\ i=1,\dots,\bar{l}\ ,$$

with some functions $\varphi_n \in C^\infty[0,b]$ and non-negative integers s depending on i,j and k such that $\varphi_n(0) = 0$ (n=0,...,s).

<u>Proof.</u> Let $\tilde{\lambda}_1$, for example, be a characteristic factor of multiplicity r and fix j and k. Substituting $x \to x^m$, we obtain from

$$x^{\varkappa_1 - \mu_{1j}} e^{-\hat{\lambda}_1} \hat{A} e^{\hat{\lambda}_1} x^{\mu_{1j}}\ , \quad \varkappa_1 = \varkappa(\hat{A}_1),$$

a differential operator B with coefficients in $C^\infty[0,b]$. Starting from a factorization (2.1.5) of $\hat{A}$ which satisfies $\tilde{\lambda}_1 = \dots = \tilde{\lambda}_r = 0$, we observe that the corresponding formal differential operator $\hat{B}$ has a factorization

$$\hat{B} = b_1(\partial - \Lambda_1)\dots(\partial - \Lambda_1),\ b_1, \Lambda_i \in \hat{\mathfrak{A}}_1,\ \ \tilde{\Lambda}_i = 0\ (i=1,\dots,r)\ .$$

Then $\hat{C} = (\partial - \Lambda_r)\dots(\partial - \Lambda_1)$ is a formal Fuchsian operator with coefficients in $\mathbb{C}[[x]]$, i.e. all characteristic factors of $\hat{C}$ vanish.

By Theorem 1.6.4, or rather its proof, there exists a formal log-power series

$$\hat{v}(x) = (\ln x)^k + \sum_{0\leq n\leq s} \hat{\varphi}_n(x)(\ln x)^n,\ \hat{\varphi}_n \in \mathbb{C}[[x]],$$

such that $\hat{\varphi}_n(0) = 0$ for all n and $\hat{C}\hat{v} = 0$. By Borel's theorem (cf. Narasimhan [1]), there exist $\psi_n \in C^\infty[0,b]$ satisfying $\psi_n^{(j)}(0) = \hat{\varphi}_n^{(j)}(0)$, $j \in \mathbb{N}_0$. Since $\hat{B}\hat{v} = 0$, we obtain

$$g = B[(\ln x)^k + \sum_{0\leq n\leq s} \psi_n(x)(\ln x)^n] \in C^{\infty,0}[0,b].$$

By Theorem 2.4.6 we find a solution $z \in C^{\infty,0}[0,b]$ of the equation $Bz = g$. Consequently, the function

$$v = (\ln x)^k + \sum_{0\leq n\leq s} \varphi_n(x)(\ln x)^n,\ \ \varphi_0 = \psi_0 - z,\ \ \varphi_n = \psi_n (n>0),$$

is a solution of $Bv = 0$ on $(0,b]$, where $\varphi_n(0) = 0$ for all n. Thus

$$x^{\mu_1 j_e \hat{\lambda}_1} v(x^{1/m})$$

is the desired solution of equation (2.5.3) on $(0,b]$.

Finally, we notice that the functions (2.5.4) are linearly independent on $(0,b]$, completing the proof of the theorem. □

2.6. A class of differential operators in weighted Sobolev spaces

Let A be the differential operator (2.0.1) and $\hat{A}$ the corresponding formal differential operator (2.3.1). Furthermore, let μ_i $(i=1,\dots,l_0)$ be the roots of the characteristic equation (2.1.12) and γ_i $(i=l_0+1,\dots,l)$ the roots of equations (2.2.3). Recall that $\gamma_i x^{1-q_i}$, with q_i defined by (2.2.2), are the lowest terms of the non-vanishing characteristic factors of A. It turns out that under the condition

$$\operatorname{Re}\gamma_i \neq 0\ ,\ i=l_0+1,\dots,l\ , \tag{2.6.1}$$

the weighted spaces X_p^{ϱ} can simply be described and the index of A depends only on the numbers μ_i and γ_i. Moreover, in this case we obtain solvability results for the operator (2.0.1) in weighted Sobolev spaces which duplicate those for Fuchsian operators in Theorem 1.4.2.

2.6.1. With A we now associate the Banach space

$$L_{p,A}^{\varrho}(0,b) = \{ y \in L_p^{\varrho}(0,b) : x^{q(i)}D^i y \in L_p^{\varrho}(0,b)\ ,\ i=1,\dots,l \}$$

endowed with the norm

$$\| y \|_{L_{p,A}^{\varrho}(0,b)} = \sum_{0\le i\le l} \| x^{q(i)}D^i y \|_{L_p^{\varrho}(0,b)}\ , \tag{2.6.2}$$

where $q(i)=q_0+\dots+q_i$, $q_0=0$ and $q_1=\dots=q_{l_0}=1$.

Lemma 2.6.1. Assume (2.6.1). Then the spaces $X_p^{\varrho}(A_0;(0,b))$ and $L_{p,A}^{\varrho}(0,b)$ coincide for any principal part A_0 of A of the form (2.3.2).

Proof. By (2.4.3) we have

$$\| y \|_{L_{p,A}^{\varrho}} \le c \| y \|_{X_p^{\varrho}}\ ,\ y \in X_p^{\varrho}\ .$$

Since $w(i) = i-q(i)$ in view of (2.6.1), we also obtain

$$\|y\|_{X_p^\varrho} = \sum_{0\le i\le l} \|B_i y\|_{L_p^{\varrho-q(i)+i}} \le c_1 \|y\|_{L^\varrho_{p,A}} \quad , \; y\in L^\varrho_{p,A} \, . \; \square$$

Moreover, under hypothesis (2.6.1), $q-l+w(l) = q-q(l) = -\varkappa$ (cf. (2.2.5)) and thus $Y_p^\varrho(0,b) = L_p^{\varrho-\varkappa}(0,b)$. Let ξ (resp. $\xi(p,\varrho)$) be the number of indices i such that $\operatorname{Re}\gamma_i > 0$ (resp. $\operatorname{Re}\mu_i > -1/p+\varrho$). As a consequence of Theorem 2.4.1 and Lemma 2.6.1 we obtain

<u>Theorem 2.6.2.</u> Under the hypotheses (2.6.1) and

$$\operatorname{Re}\mu_i \neq -1/p+\varrho\,, \; i=1,\ldots,l_0 \,, \tag{2.6.3}$$

$A\in L(L^\varrho_{p,A}(0,b),L_p^{\varrho-\varkappa}(0,b))$ is a surjective operator with kernel index $\xi+\xi(p,\varrho)$.

<u>Remark 2.6.3.</u> If we assume (2.6.1) and associate the principal part

$$\begin{aligned} &A_0 = x^{-\varkappa}A_1\ldots A_l \,,\; A_i = x^{q_i}D-c_i \,,\; c_i = \mu_i \; (i=1,\ldots,l_0)\,, \\ &c_i = \gamma_i \; (i=l_0+1,\ldots,l) \end{aligned} \tag{2.6.4}$$

with A, then relation (2.4.1) holds. Furthermore, Theorem 2.6.2 remains valid for any differential operator of the form

$$B = A_0 + \sum_{0\le i\le l} b_i(x)x^{q(i)-\varkappa}D^i \,,\; b_i\in C^0[0,b],\; b_i(0)=0, \quad i=0,\ldots,l \, .$$

As in the proof of Theorem 1.3.1(i), this follows from Theorem 2.3.1 since

$$\|B-A_0\|_{L(L^\varrho_{p,A}(0,b),L_p^{\varrho-\varkappa}(0,b))} \to 0 \quad \text{as } b\to 0 \, .$$

We now show that hypotheses (2.6.1) and (2.6.3) are even necessary in Theorem 2.6.2.

<u>Theorem 2.6.4.</u> If one of the conditions (2.6.1) and (2.6.3) is violated, then $A\in L(L^\varrho_{p,A}(0,b),L_p^{\varrho-\varkappa}(0,b))$ is not normally solvable.

<u>Proof.</u> Without loss of generality we may assume $\bar a_l=1$. For $\varepsilon\in\mathbb{R}$, consider the operators

$$A(\varepsilon) = A_0(\varepsilon)+A-A_0 \,,\; A_0(\varepsilon) = x^{-\varkappa}(A_1-\varepsilon)\ldots(A_l-\varepsilon)$$

with A_0 and A_i defined in (2.6.4). We observe that

$$\| A(\varepsilon)-A \|_{L(L^{\varrho}_{p,A}(0,b),L^{\varrho-\varkappa}_p(0,b))} \to 0 \quad \text{as} \quad \varepsilon \to 0 ,$$

so that $A(\varepsilon)$ is a small perturbation of A with respect to the operator norm. Let ξ_ε (resp. $\xi_\varepsilon(p,\varrho)$) be the number of indices i such that $\operatorname{Re}\gamma_i > -\varepsilon$ (resp. $\operatorname{Re}\mu_i > -1/p+\varrho-\varepsilon$). By Remark 2.6.3, $A(\varepsilon)$ is a Fredholm operator of $L^{\varrho}_{p,A}$ into $L^{\varrho-\varkappa}_p$ with index $\xi_\varepsilon + \xi_\varepsilon(p,\varrho)$ if

$$\operatorname{Re}\mu_i \neq -1/p+\varrho-\varepsilon \ \ (i \leq l_0) , \quad \operatorname{Re}\gamma_i \neq -\varepsilon \quad (i > l_0) .$$

Since by assumption there exists an index i such that $\operatorname{Re}\mu_i = -1/p+\varrho$ or $\operatorname{Re}\gamma_i = 0$, we obtain

$$\operatorname{ind} A(\varepsilon) = \text{const} < \operatorname{ind} A(\bar{\varepsilon}) = \text{const}, \ \varepsilon \in [-c,0), \ \ \bar{\varepsilon} \in (0,c],$$

for some sufficiently small $c > 0$. As in the proof of Theorem 1.3.1(ii) we see that $A(0) = A$ cannot be normally solvable. □

<u>2.6.2.</u> We now study the operator A in spaces of functions with higher degree of smoothness. For $i=0,\dots,l$, set $\alpha(i) = \max(0,q(i)-\varkappa)$. We introduce the Banach spaces

$$W^k_{p,A}(0,b) = \{ y \in \mathcal{D}'(0,b) : x^{\alpha(i)}D^{i+j}y \in L_p(0,b) ,$$

$$i=0,\dots,l, \ j=0,\dots,k \} , \ k \in \mathbb{N}_0 ,$$

endowed with the canonical norm

$$\| y \|_{W^k_{p,A}} = \sum_{0 \leq i \leq l} \ \sum_{0 \leq j \leq k} \| x^{\alpha(i)}D^{i+j}y \|_{L_p} . \tag{2.6.5}$$

Note that, for Fuchsian operators, $W^k_{p,A}$ coincides with the space $W^k_{p,l}$ defined in 1.1.3. Let $\mathring{W}^k_p(0,b)$ be the closed subspace of $W^k_p(0,b)$ defined in 1.1.2. Set

$$\mathring{W}^k_{p,A}(0,b) = \{ y \in W^k_{p,A}(0,b) : y^{(i)}(0) = 0 , \ i=0,\dots,k+\varkappa-1 \}$$

when $k+\varkappa > 0$ and $\mathring{W}^k_{p,A} = W^k_{p,A}$ when $k+\varkappa \leq 0$.

<u>Lemma 2.6.5.</u> For $p>1$, $A \in L(\mathring{W}^k_{p,A}(0,b),\mathring{W}^k_p(0,b))$.

<u>Proof.</u> In view of (2.2.5) A can be written

$$A = \sum_{0 \leq i \leq l} c_i(x)x^{r(i)}D^i , \ r(i) = \min \{ j \in \mathbb{N}_0 : j \geq \alpha(i) \} , \tag{2.6.6}$$

where $c_i \in C^{\infty}[0,b]$. Thus it suffices to show that

$$x^{r(i)}D^i \in L(\mathring{W}^k_{p,A},\mathring{W}^k_p) \ , \ i=0,\dots,l \ . \tag{2.6.7}$$

By induction on i, we shall prove that

$$x^{q(i)-\varkappa-s}D^{i+j-s} \in L(\mathring{W}^k_{p,A},L^{k-j}_p) \tag{2.6.8}$$

for $s=0,\dots,j$, $j=0,\dots,k$ and $i=0,\dots,l$ which implies $x^{q(i)-\varkappa}D^i \in L(\mathring{W}^k_{p,A},\mathring{W}^k_p)$. Furthermore, by Hardy's inequality (Stein [1, Appendix A.4]), for any $y \in \mathring{W}^1_p$ and $p>1$ we have

$$\| x^{-1}y \|_{L_p(0,b)} = \| x^{-1} \int_0^x y'(z)dz \|_{L_p(0,b)} \leq c(p) \| y' \|_{L_p(0,b)} \ ,$$

and by induction on m,

$$x^{-1}I \in L(\mathring{W}^m_p,\mathring{W}^{m-1}_p), \ m \in \mathbb{N}, \ p>1 \ . \tag{2.6.9}$$

Therefore, $x^{\varrho} I \in L(\mathring{W}^k_p)$ for any $\varrho \geq 0$, hence

$$x^{r(i)}D^i = x^{\varrho}\, x^{q(i)-\varkappa}D^i \in L(\mathring{W}^k_{p,A},\mathring{W}^k_p), \ \varrho = r(i)-q(i)+\varkappa, \quad \text{Q.E.D.}$$

For the proof of (2.6.8), we consider two cases. Let first $\varkappa \geq 0$. By (2.6.9) and the continuous embedding $\mathring{W}^k_{p,A} \subset \mathring{W}^k_p$, we obtain (2.6.8) for $i=0,\dots,\varkappa$. Suppose (2.6.8) holds for all $i<\bar{i}$, where $\varkappa < \bar{i} \leq l$. By assumption and the relations

$$q(\bar{i})-1 \geq q(\bar{i}-1), \ x^{\alpha(\bar{i})}D^{\bar{i}+k} \in L(W^k_{p,A},L_p) \ ,$$

we conclude (2.6.8) for $i=\bar{i}$.

Let now $\varkappa < 0$. Then $x^{-\varkappa}D^k \in L(W^k_{p,A},L_p)$. By another inequality of Hardy (cf. Stein [1, Appendix A.4])

$$\| x^{-\varkappa-1} [y^{(k-1)}(x) - y^{(k-1)}(b)] \|_{L_p(0,b)}$$
$$= \| x^{-\varkappa-1} \int_b^x z \ \ (z^{-} \ y^{(k)})dz \|_{L_p(0,b)} \leq c \, \| x^{-\varkappa} y^{(k)} \|_{L_p(0,b)}$$

for all $y \in W^k_{p,A}$. Moreover, by Sobolev's inequality

$$| y^{(k-1)}(b) | \leq c_1 \| y \|_{W^k_p(b/2,b)} \leq c_2 \| y \|_{W^k_{p,A}(0,b)}, \ y \in W^k_{p,A}$$

so that $x^{-\varkappa-1}D^{k-1} \in L(W^k_{p,A},L_p)$. By iteration, we obtain

$$x^{-\varkappa-j}D^{k-j} \in L(\mathring{W}^k_{p,A}, L_p) \ , \ j=0,\dots,\min(-\varkappa,k) \ ,$$

which proves (2.6.8) for $i=0$ and $k+\varkappa \leq 0$. To prove (2.6.8) for $i=0$ and $k+\varkappa > 0$, one may use the continuous embedding $\mathring{W}^k_{p,A} \subset \mathring{W}^{k+\varkappa}_p$ and (2.6.9). Finally, as for $\varkappa \geq 0$, one concludes (2.6.8) when $i>0$. □

Let $\mathcal{P}_k$ $(k \in \mathbb{N}_0)$ be the linear space of polynomials of degree $\leq k$ and $\mathcal{P}_k = \{0\}$ for $k<0$. Then we obviously have

$$\mathring{W}^k_p \dotplus \mathcal{P}_{k-1} = W^k_p \ , \ \mathring{W}^k_{p,A} \dotplus \mathcal{P}_{k+\varkappa-1} = W^k_{p,A} \ , \tag{2.6.10}$$

and by Lemma 2.6.5, A is a continuous map of $W^k_{p,A}$ into W^k_p.

Theorem 2.6.6. Assume $p>1$.

(i) Under the hypotheses (2.6.1) and (2.6.3) for $\varrho = k+\varkappa$, $A \in L(W^k_{p,A}(0,b), W^k_p(0,b))$ is a Fredholm operator with index

$$\operatorname{ind} A = \xi + \xi(p,k+\varkappa) - \min(k,-\varkappa) \ . \tag{2.6.11}$$

Moreover, the equation $Ay = f \in W^k_p(0,b)$ has a solution in $W^k_{p,A}(0,b)$ if and only if, for $k+\varkappa>0$, the linear system

$$(\hat{A}\hat{y})^{(j)}(0) = f^{(j)}(0) \ , \ j=0,\dots,k-1, \ y \in \mathbb{C}[[x]] \ ,$$

in the unknowns $\hat{y}(0),\dots,\hat{y}^{(k+\varkappa-1)}(0)$ is solvable, and for $k+\varkappa \leq 0$

$$f^{(j)}(0) = 0 \ , \ j=0,\dots,k-1 \ .$$

(ii) If condition (2.6.1) or (2.6.3) for $\varrho = k+\varkappa$ is violated, then $A \in L(W^k_{p,A}(0,b), W^k_p(0,b))$ is not normally solvable.

Proof. (i) We first show that $A \in L(\mathring{W}^k_{p,A}, \mathring{W}^k_p)$ is surjective with nullity $\xi + \xi(p,k+\varkappa)$. We may assume $\bar{a}_l=1$. Starting from a factorization (2.1.5) of $\hat{A}$, we deduce that the formal differential operator $\hat{A}_k = D^k\hat{A}$ corresponding to $A_k = D^kA$ can be written

$$\hat{A}_k = x^{q-l-k}(\partial - \Lambda_{l+k})\dots(\partial - \Lambda_1), \ \Lambda_i \in \hat{\mathcal{O}} \ , \tag{2.6.12}$$

where

$$\tilde{\Lambda}_{i+k} = \tilde{\lambda}_i \ (i=l_0+1,\dots,l) \ , \ \ \tilde{\Lambda}_i=0 \ (i=1,\dots,l_0+k) \ ,$$

$$\Lambda_i(0) = \mu_i (i=1,\dots,l_0), \ \ \Lambda_{i+l_0}(0) = \varkappa+i-1 \ (i=1,\dots,k).$$

Indeed, using the commutation formula (2.1.7), one obtains

$$D\hat{A} = x^{-1}\partial \hat{A} = x^{q-l-1}(\partial + q-l)(\partial - \lambda_l)\dots(\partial - \lambda_1)$$

$$= x^{q-l-1}(\partial - \lambda_l + q_l - 1 + o)(\partial + q - l + 1 - q_l + o)(\partial - \lambda_{l-1})\dots(\partial - \lambda_1)$$

$$= \dots = x^{q-l-1}(\partial - \lambda_l + q_l - 1 + o)(\partial - \lambda_{l-1} + q_{l-1} - 1 + o)\dots \times$$

$$\times(\partial - \lambda_{l_0+1} + q_{l_0+1} - 1 + o)(\partial + q - l - q_l - \dots - q_{l_0+1} + l - l_0 + o)\times$$

$$\times(\partial - \lambda_{l_0})\dots(\partial - \lambda_1),$$

where o denotes various formal Puiseux series in $\mathfrak{A}$ with vanishing zero order terms. This proves (2.6.12) for $k=1$ since $q - q_l - \dots - q_{l_0+1} = -\varkappa + l_0$. By iteration, we get (2.6.12) for $k > 1$. Moreover, we observe that $\varkappa(\hat{A}_k) = \varkappa + k$.

By virtue of (2.6.12) and (2.6.2),

$$L^{k+\varkappa}_{p,A_k}(0,b) = \{ y \in \mathcal{D}'(0,b) : x^{i-k-\varkappa} D^i y \in L_p(0,b),\ i=0,\dots,k,$$

$$x^{q(i)-\varkappa} D^{i+k} y \in L_p(0,b),\ i=1,\dots,l \},$$

and using (2.6.8), we see that

$$L^{k+\varkappa}_{p,A_k}(0,b) = \mathring{W}^k_{p,A}(0,b).$$

Applying Theorem 2.6.2 to the operator A_k, by (2.6.12) and the assumptions we obtain that $A_k \in L(\mathring{W}^k_{p,A}, L_p)$ is surjective with kernel index $\xi + \xi(p, k+\varkappa)$. From the commutative diagram

$$\begin{array}{ccc} A : \mathring{W}^k_{p,A} & \longrightarrow & \mathring{W}^k_p \\ & \searrow^{A_k} & \downarrow D^k \\ & & L_p \end{array} \qquad (2.6.13)$$

we conclude that A_k is surjective with the same kernel index since D^k is an isomorphism. Furthermore, since

$$\dim W^k_p / \mathring{W}^k_p = k, \quad \dim W^k_{p,A} / \mathring{W}^k_{p,A} = \max(0, k+\varkappa)$$

(cf. (2.6.10)), $A \in L(W^k_{p,A}, W^k_p)$ is a Fredholm operator with index

$$\xi + \xi(p,k+\varkappa) + \max(0,k+\varkappa)-k$$

which proves (2.6.11). Arguing as in the proof of Theorem 1.4.2, one finally obtains the assertion on the solvability conditions.

(ii) By (2.6.10) and (2.6.13), it is sufficient to verify that $A_k \in L(L^{k+\varkappa}_{p,A_k}, L_p)$ is not normally solvable. But this follows immediately from Theorem 2.6.4. □

Remark 2.6.7. For $\varkappa > 0$ and $l_0 > 0$, we have $\varkappa \in \{1,\dots,l_0\}$ and the characteristic equation (2.1.12) possesses the roots $0,1,\dots,\varkappa-1$ and $\gamma_i + \varkappa$ $(i=1,\dots,l_0-\varkappa)$, where γ_i are the roots of

$$\begin{aligned} &b_{l_0}(0)\gamma(\gamma-1)\dots(\gamma-l_0+\varkappa)+\dots+b_{\varkappa+1}(0)\gamma+b_{\varkappa}(0) = 0, \\ &b_i = x^{\varkappa-i}a_i . \end{aligned} \qquad (2.6.14)$$

Denoting the number of indices i satisfying $\operatorname{Re}\gamma_i > -1/p+k$ by $\eta(p,k)$, in this case we can state Theorem 2.6.6(i) as follows:

If $p>1$, $\varkappa > 0$, $\operatorname{Re}\gamma_i \neq -1/p+k$ $(i=1,\dots,l_0-\varkappa)$ and condition (2.6.1) is satisfied, then $A \in L(W^k_{p,A}(0,b), W^k_p(0,b))$ is a Fredholm operator with index $\xi + \eta(p,k) + \varkappa$.

Remark 2.6.8. Theorem 2.6.6 remains valid for any differential operator of the form

$$B = A + \sum_{0 \le i \le l} d_i(x)x^{r(i)}D^i , \quad d_i \in C^k[0,b]$$

with r(i) defined in (2.6.6) and $d_i(0) = 0$ $(i \ge \max(0,\varkappa))$ since

$$\| B-A \|_{L(\overset{\circ}{W}{}^k_{p,A}(0,b),\overset{\circ}{W}{}^k_p(0,b))} \to 0 \quad \text{as } b \to 0 .$$

Remark 2.6.9. It may be proved that the norm

$$\sum_{0 \le i \le l,\ \alpha(i) \in \mathbb{N}_0} \ \sum_{0 \le j \le k} \| x^{\alpha(i)} D^{i+j} y \|_{L_p}$$

is equivalent to (2.6.5); see Bolley, Camus and Helffer [2] for $q_i = q > 1$ $(i=1,\dots,l)$ and $p=2$.

Finally we remark that, with the help of Remark 2.3.3 and the methods of this section, one can prove an analogue of Theorem 2.6.6 for the C^k spaces (cf. Elschner and Silbermann [2]).

2.7. Examples

In this section we apply Theorem 2.6.6 and Remarks 2.6.7 and 2.6.8 to some special cases. Let $p>1$ and $k \in \mathbb{N}_0$.

Example 2.7.1. Consider the operator

$$B = a_l(x)x^{rq}D^l + a_{l-1}(x)x^{(r-1)q}D^{l-1} + \dots + a_{l-r}(x)D^{l-r} + \dots + a_0(x) ,$$

where $a_i \in C^k [0,b]$ $(i=0,\dots,l)$, $r,q \in \mathbb{N}$, $1 \leq r \leq l$ and $a_l(x) \neq 0$ for all $x \in [0,b]$. With B we associate the principal part

$$A = a_l(0)x^{rq}D^l + a_{l-1}(0)x^{(r-1)q}D^{l-1} + \dots + a_{l-r}(0)D^{l-r} .$$

For $q=1$, we have $\varkappa(A) = l-r$, $l_0(A) = l$ and the characteristic equation (2.6.14) of A takes the form

$$a_{l-r}(0) + \sum_{1 \leq i \leq r} a_{l-r+i}(0)\gamma(\gamma-1)\dots(\gamma-i+1) = 0 . \tag{2.7.1}$$

For $q>1$ and $a_{l-r}(0) \neq 0$, we have $\varkappa(A) = l_0 = l-r$ and the coefficients of the lowest terms of the non-vanishing characteristic factors coincide with the roots of the equation (cf. (2.2.3))

$$\sum_{0 \leq i \leq r} a_{l-r+i}(0)\gamma^i = 0 . \tag{2.7.2}$$

Moreover, in both cases we observe that

$$W^k_{p,A} = \{ y \in W^{k+l-r}_p(0,b) : x^{iq}D^{i+l-r}y \in W^k_p(0,b), i=1,\dots,r \} .$$

Therefore, from Theorem 2.6.6 and Remarks 2.6.7 and 2.6.8 we obtain the following results:

(i) Let γ_i be the roots of (2.7.1) and $q=1$. Then $B \in L(W^k_{p,A}, W^k_p)$ is normally solvable if and only if $\operatorname{Re}\gamma_i \neq -1/p+k$ $(i=1,\dots,r)$. If this condition is satisfied, then B has the index $\eta(p,k)+l-r$, where $\eta(p,k)$ denotes the number of indices i such that $\operatorname{Re}\gamma_i > -1/p+k$.

(ii) Let γ_i be the roots of (2.7.2), $a_{l-r}(0) \neq 0$ and $q>1$. Then $B \in L(W^k_{p,A}, W^k_p)$ is normally solvable if and only if $\operatorname{Re}\gamma_i \neq 0$ $(i=1,\dots,r)$. Under this condition, B has the index $\xi+l-r$, where ξ denotes the number of indices i such that $\operatorname{Re}\gamma_i > 0$.

Example 2.7.2. Let A be the operator (2.o.1). We assume

$q = v(a_1) = 1, \ v(a_{1-j}) \geq 1-2j \ (1 \leq j \leq r), \ r < 1/2, \ a_{1-r}(0) \neq 0,$

where v(a) denotes the order of the zero of a smooth function a at the origin. We have $\varkappa(A) = l_o(A) = 1-r$. Furthermore, since $v(a_{1-j}) > 1-jl/r$ ($j=1,\dots,r-1$), the lowest terms of the non-vanishing characteristic factors are given by $\gamma_i x^{1-l/r}$, where γ_i are the roots of the equation $a_1(0)\gamma^r = -a_{1-r}(0)$ (cf. Lemma 2.2.1). By Remark 2.6.9

$$W^k_{p,A} = \{ y \in W^{k+1-r}(0,b) : x^{il/r} D^{i+1-r} y \in W^k_p(0,b) \ , \ il/r \in \mathbb{N}_0 \ , \ 1 \leq i \leq r \} \ .$$

Now Theorem 2.6.6 yields the following results:

Under the above assumptions, $A \in L(W^k_{p,A}, W^k_p)$ is normally solvable if and only if $\operatorname{Re} \gamma_i \neq 0$ ($i=1,\dots,r$). If this condition is satisfied, then A has the index $\xi+1-r$, where ξ denotes the number of indices i such that $\operatorname{Re}\gamma_i > 0$.

<u>Example 2.7.3.</u> Consider the operator

$$A = x^q D^2 + a_1(x) x^r D + a_o(x), \ a_i \in C^k[0,b], \ a_i(0) \neq 0 \ (i=0,1) \ ,$$
$$q, r \in \mathbb{N}, \ r > 1, \ q > 2r.$$

Then, by Lemma 2.2.1, $c_1 x^{1-r}$ and $c_2 x^{1+r-q}$ with $c_1 = -a_o(0)/a_1(0)$, $c_2 = -a_1(0)$ are the lowest terms of the characteristic factors of A. Hence

$$W^k_{p,A} = \{ y \in W^k_p(0,b) : x^r Dy, x^q D^2 y \in W^k_p(0,b) \} \ .$$

Theorem 2.6.6 and Remark 2.6.8 imply that $A \in L(W^k_{p,A}, W^k_p)$ is a Fredholm operator with index ξ if $\operatorname{Re} c_i \neq 0$ ($i=1,2$). Here ξ denotes the number of indices i satisfying $\operatorname{Re} c_i > 0$.

<u>2.8. Comments and references</u>

<u>2.1.</u> The proof of Theorem 2.1.1 is due to Kuznecov [1]. The notion of determining factors already appeared in the classical literature on ordinary differential equations (cf. Cope [1], Ince [1], Sternberg [1]). Their connection with the factorization of formal differential operators was observed in Elschner [4].

<u>2.3/4.</u> Here we essentially followed Elschner [4]. The basic Theorem 2.3.2 may also be proved by estimating the norm of the integral oper-

ators (2.3.9) and (2.3.10); see Gluško [2] concerning some special cases and Elschner [4] in the general case. It was observed by Malgrange [2] that the formal differential operator (2.3.1) has index $\varkappa$ in $\mathbb{C}[[x]]$. Using Lemma II.1.3 in Deligne [1] it is possible to generalize Theorem 2.4.6 to systems of differential equations of the form

$$\mathcal{A}Y = x^q DY + A(x)Y = F \ , \ Y=(y_i)_1^l \ , \ F=(f_i)_1^l \ , \tag{2.8.1}$$

where A is an l-by-l matrix of smooth functions. The surjectivity of $\mathcal{A}$ in the space $C^{\infty,0}$ of flat vector functions was proved by Kuznecov [1] and Malgrange [2], while Elschner [2] also studied the solvability conditions for equation (2.8.1) in C^{∞}.

2.5. Theorem 2.5.1 was obtained by Svensson [1] and Theorem 2.5.2 is due to Kannai [2]. Theorem 2.5.3 is closely connected with classical results on the asymptotic behaviour of solutions to the homogeneous equation (2.5.3) with analytic coefficients; see Sternberg [1], Wasow [1] and Kannai [2].

2.6/7. Some special cases of Theorem 2.6.6(i) were obtained by Bolley, Camus and Helffer [1], [2] and Prevosto and Rolland [1] who studied differential operators with polynomial coefficients in weighted W_2^k spaces. The results of Sec. 2.6 can easily be generalized to the operator (2.0.1) on an interval $[a,b]$, $a < 0 < b$ (cf. Elschner and Silbermann [2]). Using other methods, Gluško [2] investigated the operator in Example 2.7.1. It is possible to generalize the result of Example 2.7.1 to the system (2.8.1) in the spaces W_p^k and C^k (cf. Gluško [2], Elschner [2]). For further results on degenerate systems of linear ordinary differential equations, see Natterer [1], de Hoog and Weiss [1], [3]. Moreover, under certain assumptions on the spectrum of B(0), Gluško [2] studied the operator differential equation

$$a(x)y'(x)+B(x)y(x) = f(x), \ x \in [0,b],$$

where $a(0) = 0$, $a(x) > 0$ for $x > 0$ and B(x) are continuous linear operators in a Banach space.

3. LINEAR ORDINARY DIFFERENTIAL OPERATORS WITH SEVERAL SINGULAR POINTS

In this chapter we study the index of differential operators with a finite number of singularities on an interval of the real axis which may be infinite. Using a local principle, we reduce the computation of the index to the case of one singularity. Furthermore, with the help of the substitution $x \to 1/x$, the investigation of differential operators without singularities on an infinite interval may be reduced to the results of Chap. 2. Combining these two methods, we shall show, in particular, that each differential operator with polynomial coefficients is a Fredholm operator in the Schwartz space on the real axis.

3.1. Reduction to the case of one singularity

Let $U \subset \mathbb{R}$ be an arbitrary interval, U_0 a compact subinterval of U and x_0 an interior point of U_0. The Sobolev spaces $W_p^0(U) = L_p(U)$, $W_p^k(U)$ ($k \in \mathbb{N}$) are defined as in 1.1.1 by replacing the interval $(0,b)$ by U in the definition of the norm. We set $U_1 = (-\infty, x_0] \cap U$ and $U_2 = [x_0, \infty) \cap U$. Let X_j ($j=1,2$) be given Fréchet spaces of functions defined on U_j, and $X^{(p,k)}$, $Y^{(p,k)}$, $1 \leq p \leq \infty$, $k \in \mathbb{N}_0$ (resp. $X^{(\infty)}, Y^{(\infty)}$) Fréchet spaces of functions on U such that

$$X^{(p,k)}|_{U_j} = X_j \; , \; Y^{(p,k)}|_{U_j} = Y_j \; , \; j=1,2 \; , \tag{3.1.1}$$

$$X^{(p,k)}|_{U_0} = W_p^{k+l}(U_0), \; l \in \mathbb{N} \; , \; Y^{(p,k)}|_{U_0} = W_p^k(U_0) \tag{3.1.2}$$

(resp. $X^{(\infty)}|_{U_j} = X_j \; , \; Y^{(\infty)}|_{U_j} = Y_j \; , \; j=1,2,$

$$X^{(\infty)}|_{U_0} = Y^{(\infty)}|_{U_0} = C^\infty(U_0) \quad).$$

Here the second relation of (3.1.2), for example, means that the restriction operator $y \to y|_{U_0}$ is a continuous map of $Y^{(p,k)}$ into $W_p^k(U_0)$ having a continuous right inverse. Consider a linear differential operator A of order l with coefficients in $C^\infty(U)$ and assume $A \in L(X_j, Y_j)$ ($j=1,2$).

<u>Theorem 3.1.1.</u> (i) If $A \in L(X_j, Y_j)$ ($j=1,2$) are both Fredholm operators

with indices $ind_j A$, then $A \in L(X^{(p,k)}, Y^{(p,k)})$ (resp. $A \in L(X^{(\infty)}, Y^{(\infty)})$) is a Fredholm operator with index

$$ind\, A = ind_1 A + ind_2 A - l \; . \qquad (3.1.3)$$

(ii) If one of the operators $A \in L(X_j, Y_j)$ $(j=1,2)$ is not normally solvable, then $A \in L(X^{(p,k)}, Y^{(p,k)})$ is not normally solvable, too.

<u>Proof.</u> 1. We first show assertion (i) for $A \in L(X^{(p,k)}, Y^{(p,k)})$. Define the subspaces

$$\bar{X}_j = \{ y \in X_j : y^{(r)}(x_o) = 0, r=0,\dots,k+l-1 \} \subset X_j \; ,$$

$$\bar{Y}_j = \{ y \in Y_j : y^{(r)}(x_o) = 0, r=0,\dots,k-1 \} \subset Y_j, \; j=1,2;$$

$$\bar{X}^{(p,k)} = \{ y \in X^{(p,k)} : y^{(r)}(x_o) = 0, r=0,\dots,k+l-1 \} \subset X^{(p,k)} \; ,$$

$$\bar{Y}^{(p,k)} = \{ y \in Y^{(p,k)} : y^{(r)}(x_o) = 0, r=0,\dots,k-1 \} \subset Y^{(p,k)} \; ,$$

each of them being closed in view of (3.1.2). Moreover,

$$\left. \begin{array}{l} \dim X_j/\bar{X}_j = k+l \; , \quad \dim Y_j/\bar{Y}_j = k \; , \; j=1,2 \; ; \\ \dim X^{(p,k)}/\bar{X}^{(p,k)} = k+l \; , \; \dim Y^{(p,k)}/\bar{Y}^{(p,k)} = k \; . \end{array} \right. \qquad (3.1.4)$$

Consider the commutative diagram

$$\begin{array}{llcl} A : & \bar{X}^{(p,k)} & \longrightarrow & \bar{Y}^{(p,k)} \\ & \downarrow R & & \downarrow R \\ \bar{A} : & \bar{X}_1 \times \bar{X}_2 & \longrightarrow & \bar{Y}_1 \times \bar{Y}_2 \end{array} \qquad , \qquad (3.1.5)$$

where $\bar{A}$ is defined by $\bar{A}(y_1,y_2) = (Ay_1, Ay_2)$ and $Ry = (y_{|U_1}, y_{|U_2})$. By assumption and (3.1.4), $A \in L(\bar{X}_j, \bar{Y}_j)$ are Fredholm operators with indices $ind_j A - l$. Consequently, $\bar{A}$ is a Fredholm operator with index $ind_1 A + ind_2 A - 2l$. Since R are isomorphisms, we obtain from (3.1.5) and (3.1.4) that $A \in L(X^{(p,k)}, Y^{(p,k)})$ has index (3.1.3).

2. Next, we verify (i) for $A \in L(X^{(\infty)}, Y^{(\infty)})$. We introduce the Fréchet spaces

$$X^{(k)} = \{ y : y_{|U_j} \in X_j, j=1,2, y_{|U_o} \in W_2^{k+l}(U_o) \} \; ,$$

$$Y^{(k)} = \{ y : y_{|U_j} \in Y_j, j=1,2, y_{|U_o} \in W_2^{k}(U_o) \} \; , \; k \in \mathbb{N} \; .$$

Then $X^{(\infty)}$ and $Y^{(\infty)}$ is the projective limit of the spaces $X^{(k)}$ and $Y^{(k)}$, respectively. As in 1. one can prove that $A \in L(X^{(k)},Y^{(k)})$ is a Fredholm operator with index (3.1.3) for any $k \in \mathbb{N}$, and taking $k \to \infty$, we obtain the result.

3. Finally, we prove (ii). By (3.1.4) and Lemma V.1.5 in Goldberg [1], $A \in L(X^{(p,k)},Y^{(p,k)})$ is normally solvable if and only if $A \in L(\bar{X}^{(p,k)}, \bar{Y}^{(p,k)})$ is normally solvable. By virtue of (3.1.5), the latter assertion is equivalent to the normal solvability of $A \in L(\bar{X}_j,\bar{Y}_j)$ $(j=1,2)$, which is equivalent to the normal solvability of $A \in L(X_j,Y_j)$ $(j=1,2)$ in view of (3.1.4) and the lemma cited above. Thus, by assumption, $A \in L(X^{(p,k)}, Y^{(p,k)})$ cannot be normally solvable. □

3.2. A class of selfadjoint differential operators with two singularities

3.2.1. We consider differential operators of the form

$$A = \sum_{0 \le i \le l} c_i(x)(x-x_1)^{\alpha_1(i)}(x_2-x)^{\alpha_2(i)}D^i , \tag{3.2.1}$$

where $-\infty < x_1 < x_2 < \infty$, $\alpha_s(i) = \max(0,l_s-l+i), l_s \in \{0,1,\dots,l\}$ $(s=1,2)$, $c_i \in C^{\infty}[x_1,x_2]$ $(i=0,\dots,l)$, $(-1)^{l/2}c_l(x) > 0$ $(x \in [x_1,x_2])$.

For $l_s > 0$, $x=x_s$ is a singular point of A. Let $k \in \mathbb{N}_0$ and $1 < p \le \infty$. We introduce the weighted Sobolev spaces

$$W^k_{p,A}(x_1,x_2) = \{ y \in W^k_p(x_1,x_2) : (x-x_1)^{\alpha_1(i)}(x_2-x)^{\alpha_2(i)}D^i y \in W^k_p(x_1,x_2),\ i=1,\dots,l \}$$

with the canonical norm

$$\sum_{0 \le i \le l} \| (x-x_1)^{\alpha_1(i)}(x_2-x)^{\alpha_2(i)}D^i y \|_{W^k_p(x_1,x_2)}.$$

Following Remark 2.6.7 and Example 2.7.1, with A we associate the characteristic equations

$$Q_s(\gamma) = \sum_{0 \le i \le l_s} c_{i+l-l_s}(x_s)(x_2-x_1)^{\alpha_s(i+l-l_s)}\gamma(\gamma-1)\dots(\gamma-i+1) = 0, \quad s=1,2, \tag{3.2.2}$$

where $\alpha_{2+1}(i) = \alpha_1(i)$. Note that

$$\gamma(\gamma-1)\dots(\gamma-l+l_s+1)Q_s(\gamma+l-l_s)$$
$$= [(-1)^{s-1}(x-x_s)]^{-\gamma+l-l_s}A[(-1)^{s-1}(x-x_s)]^{\gamma}\big|_{x=x_s} ; \qquad (3.2.3)$$

see (2.1.12) and Remark 2.6.7. Denote the roots of (3.2.2) by γ_{si} $(i=1,\dots,l_s)$, and assume

$$\operatorname{Re}\gamma_{si} \neq -1/p+k, \quad i=1,\dots,l_s, \quad s=1,2. \qquad (3.2.4)$$

Let $\eta_s(p,k)$ be the number of roots satisfying $\operatorname{Re}\gamma_{si} > -1/p+k$. For $l_s=0$, condition (3.2.4) is void and we set $\eta_s=0$ in this case. As a consequence of Example 2.7.1 (for q=1) and Theorem 3.1.1 (with $X^{(p,k)} = W^k_{p,A}(U), Y^{(p,k)} = W^k_p(U), U=(x_1,x_2)$), we now obtain

<u>Theorem 3.2.1.</u> Under hypothesis (3.2.4), $A \in L(W^k_{p,A}(x_1,x_2), W^k_p(x_1,x_2))$ is a Fredholm operator with index

$$\operatorname{ind}_{p,k}A = \eta_1(p,k) + \eta_2(p,k) + l-l_1-l_2,$$

and it is not normally solvable if (3.2.4) is violated.

Furthermore, we consider boundary value problems of the form

$$Ay=f, \quad F_h(y) = \sum_{s=1,2}\ \sum_{0\le j< l-l_s} \beta^s_{hj}y^{(j)}(x_s) = 0, \quad h=1,\dots,n, \qquad (3.2.5)$$

where $\beta^s_{hj} \in \mathbb{C}$ and $0 \le n \le 2l-l_1-l_2$. Henceforth we shall assume that the rows of the matrix

$$(\beta^1_{h1}\dots\beta^1_{hl-l_1-1}\ \beta^2_{h1}\dots\beta^2_{hl-l_2-1})^n_{h=1}$$

are linearly independent. Then the F_h generate linearly independent continuous linear functionals over $W^k_{p,A}$, and setting

$$\bar{W}^k_{p,A}(x_1,x_2) = \{ y \in W^k_{p,A}(x_1,x_2) : F_h(y) = 0, h=1,\dots,n \} ,$$

we obtain the relation

$$\dim W^k_{p,A} / \bar{W}^k_{p,A} = n . \qquad (3.2.6)$$

The following theorem follows immediately from (3.2.6) and Theorem 3.2.1.

Theorem 3.2.2. Under assumption (3.2.4), $A \in L(W^k_{p,A}(x_1,x_2), W^k_p(x_1,x_2))$ is a Fredholm operator with index $\mathrm{ind}_{p,k}A-n$, and it is not normally solvable if (3.2.4) is violated.

3.2.2. We now study selfadjointness and spectrum of the boundary problem (3.2.5). Consider the operator

$$\mathcal{A}y = Ay,\ D(\mathcal{A}) = C^\infty_{\mathcal{A}}[x_1,x_2] = \{ y \in C^\infty[x_1,x_2] \ : \ F_h y = 0,\ h=1,\dots,n\} \tag{3.2.7}$$

which is generated by (3.2.5). If n=0, we set $D(\mathcal{A}) = C^\infty[x_1,x_2]$. (3.2.7) is called formally selfadjoint in $L_2(x_1,x_2)$ if

$$(Ay,v) = (y,Av)\ ,\ y,\ v \in C^\infty_{\mathcal{A}}[x_1,x_2]\ , \tag{3.2.8}$$

where (.,.) denotes the scalar product in the (complex) Hilbert space $L_2(x_1,x_2)$. (3.2.8) implies that (3.2.1) is formally selfadjoint in the usual sense:

$$A = A^* = \sum_{0 \le i \le 1} (-1)^i D^i (x-x_1)^{\alpha_1(i)} (x_2-x)^{\alpha_2(i)} \overline{c_i(x)}\ . \tag{3.2.9}$$

In particular, the generalized Legendre operators

$$\mathcal{B}y = By,\ D(\mathcal{B}) = \{ y \in C^\infty[x_1,x_2] : y^{(j)}(x_s) = 0,\ j=0,\dots,l-m-1,$$

$$s=1,2\}\ ,\ B = (-1)^l D^l (a^m) D^l, m \in \{1,\dots,2l-1\}\ , \tag{3.2.10}$$

$$a(x) = (x-x_1)(x_2-x)b(x),\ b \in C^\infty[x_1,x_2]\ ,\ b(x) > 0\ (x \in [x_1,x_2]\ )$$

are formally selfadjoint operators of the form (3.2.7). For $m \ge l$, we set $D(\mathcal{B}) = C^\infty[x_1,x_2]$.

The operator (3.2.7) is called essentially selfadjoint if its closure $\bar{\mathcal{A}}$ in $L_2(x_1,x_2)$ is selfadjoint. In the sequel we need the following

Lemma 3.2.3. $C^\infty_{\mathcal{A}}[x_1,x_2]$ is dense in $\overline{W}^0_{2,A}(x_1,x_2)$.

Proof. We first show that C^∞ is dense in $W^0_{2,A}$. It is sufficient to consider the case when A has only one singularity. Let $l_2 = 0$ and $x_1 = 0$, $x_2 = 1$, for example. Then we have

$$W^0_{2,A} = \{ y \in W_2^{l-l_1}(0,1)\ :\ x^i D^{i+l-l_1} y \in L_2(0,1), i=1,\dots,l_1 \}\ .$$

Let first $l_1=l$. We choose a function $\chi \in C^{\infty}[0,1]$ such that $0 \le \chi \le 1$ for $x \in [0,1]$, $\chi=0$ for $x \le 1/3$ and $\chi=1$ for $x \ge 2/3$, and approximate an arbitrary element $y \in W^{o}_{2,A}$ by the sequence $y_n = \chi(nx)y$. Indeed, $y_n \in W^{l}_{2}$ for all $n \in \mathbb{N}$, and using the relations

$$x^iD^i(y_n-y) = (\chi(nx)-1)x^iD^iy$$

$$+ x^i \sum_{1 \le j \le i} \binom{i}{j} n^j \chi^{(j)}(nx)D^{i-j}y, \quad i=0,\dots,l,$$

$$\| (\chi(nx)-1)x^iD^iy \|_{L_2} \to 0, \quad \| x^in^j\chi^{(j)}(nx)D^{i-j}y\|_{L_2} \to 0,$$

$$n \to \infty \quad , \; j=1,\dots,i, \; i=0,\dots,l \; ,$$

we obtain $y_n \to y$ in $W^{o}_{2,A}$. Thus W^{l}_{2} is dense in $W^{o}_{2,A}$ when $l_1=l$. For $l_1 < l$ and $y \in W^{o}_{2,A}$, set $z = D^{l-l_1}y$ and approximate z as above by $z_n = \chi(nx)z$. With

$$y_n = J^{l-l_1}z_n + \sum_{0 \le i < l-l_1} y^{(i)}(0)x^i/i!, \quad Jz = \int_0^x z(t)dt$$

we get $y_n \in W^{l}_{2}$ and $y_n \to y (n \to \infty)$ in $W^{o}_{2,A}$ again. Since C^{∞} is dense in W^{l}_{2}, it is also dense in $W^{o}_{2,A}$.

To complete the proof of the lemma, we observe that in view of (3.2.6) and Lemma 2.3 in Przeworska-Rolewicz and Rolewicz [1, Chap. B.III] $C^{\infty}_{\mathcal{A}} = C^{\infty} \cap \overline{W}^{o}_{2,A}$ is dense in $\overline{W}^{o}_{2,A}$. □

We say that the spectrum of a linear operator A in a Hilbert space H is discrete if it consists entirely of isolated eigenvalues of finite multiplicity. Following Goldberg [1], the essential spectrum of A is defined to be the set of all $\lambda \in \mathbb{C}$ such that $A-\lambda I$ is not normally solvable in H. We are now ready to prove the main result of this section.

<u>Theorem 3.2.4.</u> Let the operator $\mathcal{A}$ defined in (3.2.7) be formally selfadjoint. $\mathcal{A}$ is essentially selfadjoint and $D(\bar{\mathcal{A}}) = \overline{W}^{o}_{2,A}(x_1,x_2)$ if and only if one of the following conditions is satisfied:

(i) $l_1=l_2=l$, $l=2m$, $m \in \mathbb{N}$;

(ii) $l_1=l$, $l_2<l$, $l=2m$, $m \in \mathbb{N}$, $\operatorname{Re} \gamma_{2i} \neq -1/2$ $(i=1,\dots,l_2)$,

$\eta_2(2,0)+m-l_2 = n$ (the case $l_1<l$, $l_2=l$ is analogous);

(iii) $l_s < l$, $\operatorname{Re}\gamma_{si} \neq -1/2$ $(i=1,\dots,l_s,\ s=1,2)$, $\eta_1(2,0)+\eta_2(2,0)$ $+ l-l_1-l_2 = n$.

Moreover, under condition (iii), the spectrum of $\bar{A}$ is discrete. The essential spectrum of $\bar{A}$ coincides with the interval $[M,\infty)$ in case (i) and with $[M_1,\infty)$ in case (ii), where $M = \min(M_1,M_2)$ and $M_s = \inf_{z\in\mathbb{R}} Q_s(-1/2+z\sqrt{-1})$.

Proof. Since C^∞ is dense in $\bar{W}^0_{2,A}$ by Lemma 3.2.3, the operator $A, D(A) = \bar{W}^0_{2,A}(x_1,x_2)$ is symmetric in $L_2(x_1,x_2)$. It is selfadjoint if and only if $A-\lambda I \in L(\bar{W}^0_{2,A}(x_1,x_2), L_2(x_1,x_2))$ is a Fredholm operator with index 0 for all $\lambda\in\mathbb{C}$ such that $\operatorname{Im}\lambda \neq 0$. Moreover, then we also obtain $D(\bar{A}) = \bar{W}^0_{2,A}(x_1,x_2)$ since $\bar{A}$ is a symmetric extension of the operator $A, D(A) = \bar{W}^0_{2,A}$ in view of Lemma 3.2.3. We now apply Theorem 3.2.2 to the operators $A-\lambda I$.

1. Let $l_1=l_2=l$. $A-\lambda I$ has the characteristic equations $Q_s(\gamma)-\lambda=0$ $(s=1,2)$. (3.2.3) and (3.2.9) imply $\overline{Q_s(\gamma)}=Q_s(-\bar{\gamma}-1)$. Consequently, the polynomial $R_s(z) = Q_s(-1/2+z\sqrt{-1})$ has real coefficients and its leading coefficient is positive. Now it follows from Gantmacher [1, Chap. 15] that, for $\operatorname{Im}\lambda\neq 0$, the number of roots of $R_s(z)=\lambda$ lying in the half-plane $\operatorname{Im} z>0$ is given by

$$n_s(\lambda) = \{\arg(R_s(z)-\lambda)\}_{z=-\infty}^{\infty} + 1/2 ,$$

where $\{\arg\cdot\}$ denotes the variation of the argument. Therefore we obtain $n_s(\lambda) = m$ when $l=2m$ and $n_s(\lambda) = m+1$ for $\operatorname{Im}\lambda > 0$ (resp. m for $\operatorname{Im}\lambda < 0$) when $l=2m+1$. By Theorem 3.2.2, $A-\lambda I \in L(W^0_{2,A}, L_2)$, $\operatorname{Im}\lambda \neq 0$, are Fredholm operators with index 0 if and only if l is even. Furthermore, Theorem 3.2.2 implies that $\bar{A}-\lambda I$ is not normally solvable if and only if

$$\lambda \in \{Q_s(-1/2+z\sqrt{-1}) : z\in\mathbb{R},\ s=1,2\} = [M,\infty) .$$

Thus the essential spectrum of $\bar{A}$ is $[M,\infty)$. The theorem is proved for $l_1=l_2=l$.

2. Let $l_s<l$ $(s=1,2)$. In case (iii), $A-\lambda I\in L(\bar{W}^0_{2,A}, L_2)$ is a Fredholm operator with index 0 for all λ. Hence the spectrum of $\bar{A}$ is discrete.

Conversely, if $\eta_1(2,0)+\eta_2(2,0)+l-l_1-l_2 \neq n$ or $\mathrm{Re}\,\gamma_{si} = -1/2$ for some root γ_{si} of (3.2.2), then by Theorem 3.2.2, there is no $\lambda \in \mathbb{C}$ such that $A-\lambda I \in L(\bar{W}^{o}_{2,A}, L_2)$ is Fredholm with index 0.

3. Let $l_1 = l$ and $l_2 < l$. Using the above arguments, one can show that $A-\lambda I \in L(\bar{W}^{o}_{2,A}, L_2)$, $\mathrm{Im}\,\lambda \neq 0$, are all Fredholm operators with index 0 if and only if (ii) is valid. Similarly to 1. one obtains the essential spectrum of $\bar{\mathcal{A}}$. □

<u>Remark 3.2.5.</u> If $l_1=l_2=2m$, then n=0 and $\mathcal{A}$=A and each formally selfadjoint operator (3.2.1) is essentially selfadjoint.

Theorem 2.2.4 can be used to study the invertibility of a selfadjoint boundary value problem.

<u>Theorem 3.2.6.</u> Let $2 \leq p < \infty$ and suppose the formally selfadjoint operator $\mathcal{A}$ satisfies condition (iii) of Theorem 3.2.4. Let $\sigma(\bar{\mathcal{A}})$ be the spectrum of $\bar{\mathcal{A}}$. Then $A-\lambda I \in L(\bar{W}^{k}_{p,A}(x_1,x_2), W^{k}_{p}(x_1,x_2))$ is invertible for all $\lambda \bar{\in} \sigma(\bar{\mathcal{A}})$ if and only if

$$\mathrm{Re}\,\gamma_{si} \bar{\in} (-1/2, -1/p+k] \ , \ i=1,\dots,l_s, \ s=1,2. \qquad (3.2.11)$$

<u>Proof.</u> By Theorem 3.2.4 the operator $A-\lambda I$, $\lambda \bar{\in} \sigma(\bar{\mathcal{A}})$, has trivial kernel in $\bar{W}^{k}_{p,A}$, and by (3.2.11) and Theorem 3.2.2, it is an invertible map of $\bar{W}^{k}_{p,A}$ onto W^{k}_{p}. Conversely, if (3.2.11) is violated, then for any $\lambda \in \mathbb{C}$, $A-\lambda I \in L(\bar{W}^{k}_{p,A}, W^{k}_{p})$ is not a Fredholm operator with index 0 in view of Theorem 3.2.2. □

<u>3.2.3.</u> As an example we study the generalized Legendre operators (3.2.10). The characteristic equations (3.2.2) take the form

$$(-1)^{l} b(x_s)^{m}(x_2-x_1)^{m}(\gamma+l)(\gamma+l-1)\dots(\gamma+l-m+1) = 0 \ .$$

Their roots are $-l, -l+1, \dots, -l+m-1$, so condition (3.2.4) is satisfied for all $p \in (1,\infty)$ and $k \in \mathbb{N}_0$. Moreover, we have

$$\eta_s(p,k) = \max(0, m-l-k) \ (s=1,2), \ n=2\max(0, l-m).$$

Furthermore, for all $y,v \in D(\mathcal{B})$,

$$(By,v) = (y,Bv) \ , \ (By,y) \geq 0 \ .$$

Consequently, $\sigma(\bar{\mathcal{B}}) \subset [0,\infty)$. The space $\bar{W}^{k}_{p,B}$ can be described as follows:

$$\bar{W}^k_{p,B}(x_1,x_2) = \{ y \in W_p^{k+2l-m}(x_1,x_2) : (x-x_1)^i (x_2-x)^i D^{i+2l-m} y \in W_p^k(x_1,x_2), i=1,\dots,m; y^{(j)}(x_s)=0, j=0,\dots,l-m-1, s=1,2 \}.$$

As a consequence of Theorems 3.2.4 and 3.2.6, we now obtain

Theorem 3.2.7. $\mathfrak{B}$ is essentially selfadjoint, $D(\bar{\mathfrak{B}}) = \bar{W}^0_{2,B}(x_1,x_2)$ and the spectrum of $\bar{\mathfrak{B}}$ is discrete. Moreover, $B-\lambda I \in L(\bar{W}^k_{p,B}(x_1,x_2), W_p^k(x_1,x_2))$ $(2 \le p < \infty)$ is invertible for all $\lambda \bar{\in} [0,\infty)$ if and only if either $m \le l$ or $m > l$ and $k=0$.

3.3. Differential operators without singularities on an infinite interval

3.3.1. For an infinite interval $U \subset \mathbb{R}$, let $S(U)$ be the Schwartz space of infinitely differentiable rapidly decreasing functions on U. $S(U)$ endowed with the system of seminorms

$$\|y\|_{k,n} = \sup_{x \in U} \{(1+|x|)^n |y^{(k)}(x)|\} \quad , \; k,n \in \mathbb{N}_0$$

is a Fréchet space. Let further $\bar{C}^\infty(U)$ be the set of all functions $y \in C^\infty(U)$ having an asymptotic expansion

$$y(x) \sim \sum_{j \ge 0} c_j x^{N-j} \quad \text{as } x \to +\infty, \; c_j \in \mathbb{C}, \; N \in \mathbb{Z} \tag{3.3.1}$$

when $\sup U = +\infty$ and

$$y(x) \sim \sum_{j \ge 0} d_j x^{M-j} \quad \text{as } x \to -\infty, \; d_j \in \mathbb{C}, \; M \in \mathbb{Z} \tag{3.3.2}$$

when $\inf U = -\infty$, which can be differentiated up to an arbitrary order:

$$y^{(i)}(x) \sim \sum_{j \ge 0} c_j (N-j)\dots(N-j+1-i) x^{N-j-i} \quad \text{as } x \to +\infty,$$

$$y^{(i)}(x) \sim \sum_{j \ge 0} d_j (M-j)\dots(M-j+1-i) x^{M-j-i} \quad \text{as } x \to -\infty, \; i \in \mathbb{N}_0.$$

Here (3.3.1), for example, is to be interpreted as

$$y(x) - \sum_{0 \le j \le n} c_j x^{N-j} = O(x^{N-n-1}) \; (x \to \infty) \text{ for all } n \in \mathbb{N}_0 .$$

In analogy to Chap. 2.1 we now introduce some definitions which will be used for the classification of differential operators on an infinite interval. For a function $y \in \bar{C}^\infty[a,\infty)$ with the asymptotic expansion (3.3.1), set $v_\infty(y) = N$ when $c_0 \ne 0$ and $v_\infty(y) = -\infty$ when $c_j=0$ for all

$j \in \mathbb{N}_0$. We consider differential operators of the form

$$A = \sum_{0 \le i \le l} a_i(x) D^i, a_i \in \bar{C}^{\infty}[a,\infty) \ (i=0,\dots,l),\ a>0,\ a_l(x) \neq 0$$

$$\text{for } x \in [a,\infty),\ v_\infty(a_l) > -\infty . \tag{3.3.3}$$

The integer $l_\infty = l_\infty(A)$ is specified by the requirements

$$\begin{aligned} l_\infty - v_\infty(a_{l_\infty}) &\le i - v_\infty(a_i) \quad \text{for } i < l_\infty , \\ l_\infty - v_\infty(a_{l_\infty}) &< i - v_\infty(a_i) \quad \text{for } i > l_\infty . \end{aligned} \tag{3.3.4}$$

Furthermore, we define

$$\varkappa_\infty = \varkappa_\infty(A) = l_\infty - v_\infty(a_{l_\infty}). \tag{3.3.5}$$

With (3.3.3) we associate the formal differential operator

$$\hat{A}_\infty = \sum_{0 \le i \le l} \hat{a}_{\infty i} D^i ,$$

where $\hat{a}_{\infty i}$ denote the asymptotic expansions of a_i as $x \to \infty$. Substituting $x \to 1/x$ in $\hat{A}_\infty$, we obtain the formal differential operator

$$\hat{A} = \sum_{0 \le i \le l} \hat{a}_{\infty i}(1/x)(-x^2 D)^i . \tag{3.3.6}$$

By Corollary 2.1.2, (3.3.6) has a factorization

$$\hat{A} = \hat{a}_{\infty l}(1/x)(-x)^l(\partial - \lambda_l(x))\dots(\partial - \lambda_1(x)) \tag{3.3.7}$$

with certain formal Puiseux series λ_i. Let $\tilde{\lambda}_i$ be the characteristic factors of $\hat{A}$. Then the functions $\tilde{\lambda}_{\infty i}(x) = \tilde{\lambda}_i(1/x)$ $(i=1,\dots,l)$ are called the characteristic factors of A at $x=\infty$. By definitions (2.1.3), (2.1.4), (3.3.4) and (3.3.5), we have

$$l_\infty = l_0 = l_0(\hat{A}),\ \varkappa_\infty = -\varkappa = -\varkappa(\hat{A}) , \tag{3.3.8}$$

since the coefficient of D^{l-i} in $\hat{A}$ is a linear combination of the terms $a_{l-j}(1/x)x^{2l-j-i}$ $(j=0,\dots,i)$. Consequently, by Corollary 2.1.5 exactly l_∞ characteristic factors of A at ∞ vanish identically. Let ζ_∞ be the number of characteristic factors $\tilde{\lambda}_{\infty i}$ satisfying $\operatorname{Re} \tilde{\lambda}_{\infty i}(x) \to +\infty$ as $x \to +\infty$. Now we can state

<u>Theorem 3.3.1.</u> $A \in L(S[a,\infty))$ is a surjective Fredholm operator with index ζ_∞.

Proof. The map $\gamma : y(x) \to y(1/x)$ is an isomorphism of $S[a,\infty)$ onto

$$C^{\infty,0}[0,1/a] = \{v \in C^{\infty}[0,1/a] \; : \; v^{(j)}(0) = 0 \; , \; j \in \mathbb{N}_0\} \; .$$

Moreover,

$$B = \gamma A \gamma^{-1} = \sum_{0 \le i \le l} a_i(1/x)(-x^2 D)^i \; , \tag{3.3.9}$$

and after multiplication by x^r for some $r \in \mathbb{N}_0$, the coefficients of B belong to $C^{\infty}[0,1/a]$. By Theorem 2.4.6, or rather its proof, and the definition of $\tilde{\lambda}_{\infty i}$, $B \in L(C^{\infty,0}[0,1/a])$ is a surjective operator with kernel index ξ_∞. Relation (3.3.9) then completes the proof of the theorem. □

3.3.2. In analogy to 2.6.2 we now study a class of differential operators of the form (3.3.3) in weighted Sobolev spaces. For a non-vanishing characteristic factor $\tilde{\lambda}_{\infty i}$ of A, let $\gamma_i x^{1-q_i}$, $q_i < 1$, be its leading term. By Lemma 2.1.3 applied to the operator (3.3.6), the characteristic factors $\tilde{\lambda}_{\infty i}$ can be labelled in such a way that

$$1 > q_i \ge q_{i+1} \; , \; i = l_\infty + 1, \dots, l \; . \tag{3.3.10}$$

We shall assume from now on, without explicitly mentioning it, that (3.3.10) holds. Let further γ_i be the zero order terms of the formal series λ_i $(i=1,\dots,l_\infty)$ in (3.3.7). Applying Lemmas 2.1.8 and 2.2.1 to (3.3.6) and using (3.3.8), we obtain

Lemma 3.3.2. If $l_\infty < l$, then for some index s $(1 \le s \le l)$,

$$q_i = \max_{k \ge l(j)+1} \frac{v_\infty(a_k) - v_\infty(a_{l(j)})}{k - l(j)} \; , \; i = l(j)+1, \dots, l(j+1) \; , \tag{3.3.11}$$

where

$$l(j+1) = \max\{i \; : \; l(j)+1 \le i \le l, v_\infty(a_i) - v_\infty(a_{l(j)})$$

$$= q_{l(j)+1}(i - l(j))\} \; , \; j = 0, \dots, s-1, \; l(0) = l_\infty, \; l(s) = l \; .$$

Moreover, the numbers γ_i $(i=1,\dots,l_\infty)$ coincide with the roots of the characteristic equation of A (at ∞)

$$\begin{aligned} &\sum_{0 \le i \le l_\infty} b_i(-\gamma)(-\gamma-1)\dots(-\gamma-i-1) = 0 \; , \\ &b_i = \lim_{x \to \infty} x^{\varkappa_\infty - i} a_i(x) \; , \end{aligned} \tag{3.3.12}$$

while the γ_i $(i=l(j)+1,\dots,l(j+1))$ coincide with the roots of

$$\sum_{0\le i\le l(j+1)-l(j)} b_{l(j)+i}(-\gamma)^i = 0 , \tag{3.3.13}$$

$$b_{l(j)+i} = \lim_{x\to\infty} x^{-\varkappa(i,j)} a_{l(j)+i}(x),\quad \varkappa(i,j) = v_\infty(a_{l(j)}) + iq_{l(j+1)}$$

Let $\beta(i) = q_0+\dots+q_i-\varkappa_\infty$, where q_i $(i>l_\infty)$ are the numbers defined in (3.3.11), $q_0 = 0$ and $q_i = 1$ $(1\le i\le l_\infty)$. For $1\le p\le\infty$ and $k\in\mathbb{N}_0$, we introduce the Banach space

$$W^k_{p,A}(a,\infty) = \{ y\in\mathcal{D}'(a,\infty) : x^{\beta(i)}D^{i+j}y\in L_p(a,\infty), i=0,\dots,l, \\ j=0,\dots,k \}$$

with the canonical norm

$$\sum_{0\le i\le l}\ \sum_{0\le j\le k} \| x^{\beta(i)}D^{i+j}y \|_{L_p(a,\infty)} . \tag{3.3.14}$$

We have $A\in L(W^k_{p,A}(a,\infty), W^k_p(a,\infty))$ since (3.3.3) can be written

$$A = \sum_{0\le i\le l} d_i(x)x^{r(i)}D^i,\ r(i) = \max\{ j\in\mathbb{Z} : j\le\beta(i)\} ,$$

$$d_i\in\bar{C}^\infty[a,\infty),\ v_\infty(d_i)\le 0\ (i=0,\dots,l) ;$$

see Lemma 3.3.2. For the roots of (3.3.12) and (3.3.13), we assume

$$\mathrm{Re}\,\gamma_i \ne 1/p-\varkappa_\infty\ (i=1,\dots,l_\infty),\ \mathrm{Re}\,\gamma_i \ne 0\ (i=l_\infty+1,\dots,l) \tag{3.3.15}$$

with the convention $1/p = 0$ for $p = \infty$. Then the number ξ_∞ defined in 3.3.1 coincides with the number of γ_i's $(i>l_\infty)$ satisfying $\mathrm{Re}\,\gamma_i>0$. Let further $\xi_\infty(p)$ be the number of roots of (3.3.12) satisfying $\mathrm{Re}\,\gamma_i>1/p-\varkappa_\infty$.

<u>Theorem 3.3.3.</u> (i) Under hypothesis (3.3.15), $A\in L(W^k_{p,A}(a,\infty),W^k_p(a,\infty))$ is a surjective Fredholm operator with index

$$\mathrm{ind}\,A = \xi_\infty + \xi_\infty(p) . \tag{3.3.16}$$

(ii) If condition (3.3.15) is violated, then $A\in L(W^k_{p,A}(a,\infty),W^k_p(a,\infty))$ is not normally solvable.

<u>Proof.</u> 1. Let first $k=0$. The map $\psi : y(x)\to y(1/x)$ is an isomorphism of $L_p(a,\infty)$ onto $L^{2/p}_p(0,1/a) = x^{2/p}L_p(0,1/a)$. Consider the operator B

defined by (3.3.9) and the corresponding formal differential operator $\hat{A}$ defined by (3.3.6). Without loss of generality, we may assume in the following that the coefficients of B belong to $C^{\infty}[0,1/a]$. By virtue of (3.3.8) and the definition of the characteristic factors $\tilde{\lambda}_{\infty i}$, the Banach space $L^{2/p+\varkappa}_{p,B}$ introduced in 2.6.1 can be described as follows:

$$L^{2/p+\varkappa}_{p,B}(0,1/a) = \{y \in \mathcal{D}'(0,1/a) : x^{2i-\beta(i)}D^i y \in L^{2/p}_p(0,1/a),$$
$$i=0,\dots,l\}.$$

Ψ extends to an isomorphism of $\mathcal{D}'(a,\infty)$ onto $\mathcal{D}'(0,1/a)$ in a natural way and

$$\Psi x^{\beta(i)}D^i\Psi^{-1}y = x^{-\beta(i)}(-x^2D)^i y , \; y \in \mathcal{D}'(0,1/a).$$

Thus Ψ is an isomorphism of $W^{o}_{p,A}(a,\infty)$ onto $L^{2/p+\varkappa}_{p,B}(0,1/a)$, too. We observe that γ_i are the roots of the equations (2.1.12) and (2.2.3) which correspond to the operator (3.3.6). Consequently, by Theorems 2.6.2 and 2.6.4, $B \in L(L^{2/p+\varkappa}_{p,B}, L^{2/p}_p)$ is a Fredholm operator with index (3.3.16) if (3.3.15) holds, and it is not normally solvable if (3.3.15) is violated. Hence the theorem is proved for $k=0$.

2. Let $k \in \mathbb{N}$. Next, we verify that $(-D+1)^k$ is an isomorphism of $W^k_p(a,\infty)$ onto $L_p(a,\infty)$. Consider the commutative diagram

$$\begin{array}{rccc} (-D+1)^k : & W^k_p(a,\infty) & \longrightarrow & L_p(a,\infty) \\ & \downarrow \Psi & & \downarrow \Psi \\ C : & L^{2/p}_{p,C}(0,1/a) & \longrightarrow & L^{2/p}_p(0,1/a) \end{array} , \qquad (3.3.17)$$

where $C = (x^2D+1)^k$ and

$$L^{2/p}_{p,C} = \{ y \in L^{2/p}_p : x^{2i}D^i y \in L^{2/p}_p , \; i=1,\dots,k \}$$

is the space defined in 2.6.1 which corresponds to C. Since the operators Ψ in (3.3.17) are isomorphisms, we obtain the assertion from Theorem 2.6.2 applied to C.

We now consider the operator $E = CB$ on $(0,1/a)$. Using the commutation formula (2.1.7), from (3.3.7) we obtain a factorization of the corresponding formal differential operator $\hat{E} = C\hat{A}$ in which the lowest terms of the formal Puiseux series are given by

$\gamma_i x^{q_i-1}$ $(i=1,\dots,\bar{i})$, x^{-1} (k times) ,

$\gamma_i x^{q_i-1}$ $(i=\bar{i}+1,\dots,l)$,

where $\bar{i}$ is the largest index i such that $q_i > 0$ (resp. $\bar{i} = 0$ when $q_i \leq 0$ for all i). Moreover, $\varkappa(\hat{E}) = \varkappa = -\varkappa_\infty$. Consequently, the space $L^{2/p+\varkappa}_{p,E}$ defined in 2.6.1 can be described as follows:

$$L^{2/p+\varkappa}_{p,E}(0,1/a) = \{ y \in \mathcal{D}'(0,1/a) : x^{2i-\beta(i)}D^i y \in L^{2/p}_p(0,1/a),$$

$$i=0,\dots,\bar{i},\ x^{2i-\beta(\bar{i})}D^i y \in L^{2/p}_p(0,1/a),\ i=\bar{i}+1,\dots,\bar{i}+k,$$

$$x^{2(k+i)-\beta(i)}D^{i+k}y \in L^{2/p}_p(0,1/a),\ i=\bar{i}+1,\dots,l \} .$$

We set $F = (-D+1)^k A$ and consider the commutative diagram

$$\begin{array}{ccc} F : W^k_{p,A}(a,\infty) & \longrightarrow & L_p(a,\infty) \\ \downarrow \Upsilon & & \downarrow \Upsilon \\ E : L^{2/p+\varkappa}_{p,E}(0,1/a) & \longrightarrow & L^{2/p}_p(0,1/a) . \end{array} \tag{3.3.18}$$

By Theorem 2.6.2, E is a surjective Fredholm operator with index (3.3.16). Furthermore, it is easily seen that Υ are isomorphisms in (3.3.18). Thus assertion (i) is a consequence of (3.3.17) and (3.3.18). (ii) follows from (3.3.17), (3.3.18) and Theorem 2.6.4 applied to E. □

Remark 3.3.4. In analogy to Remark 2.6.9 one may prove that the norm

$$\sum_{0\leq i\leq l,\ \beta(i)\in\mathbb{Z}} \quad \sum_{0\leq j\leq k} \| x^{\beta(i)} D^{i+j} y \|_{L_p(a,\infty)}$$

is equivalent to (3.3.14).

3.4. Differential operators with one singularity on $(0,\infty)$

3.4.1. We consider differential operators of the form

$$A = \sum_{0\leq i\leq l} a_i(x) D^i,\ a_i \in \bar{C}^\infty[0,\infty)\ (i=0,\dots,l),\ a_l(x) \neq 0$$
$$(x \in (0,\infty)),\ v(a_l) < \infty ,\ v_\infty(a_l) > -\infty , \tag{3.4.1}$$

where v(a) denotes the order of the zero of a smooth function a at the origin. As in 2.6.2 and 3.3.2 we define the spaces $W^k_{p,A}(0,b)$ and

$W^k_{p,A}(c,\infty)$, where $1 < p \leq \infty$, $k \in \mathbb{N}_0$ and $0 < c < b < \infty$. We now introduce the Banach space

$$W^k_{p,A}(0,\infty) = \{ y \in \mathcal{D}'(0,\infty) : y|_{(0,b)} \in W^k_{p,A}(0,b), y|_{(c,\infty)} \in W^k_{p,A}(c,\infty) \}$$

with norm

$$\| y \|_{W^k_{p,A}(0,b)} + \| y \|_{W^k_{p,A}(c,\infty)} .$$

$W^k_{p,A}(0,\infty)$ obviously does not depend on the choice of b and c. Then Theorems 2.6.6, 3.3.3 and 3.1.1 (with $X^{(p,k)} = W^k_{p,A}(U)$, $Y^{(p,k)} = W^k_p(U)$, $U = (0,\infty)$) imply the following result for (3.4.1).

<u>Theorem 3.4.1.</u> Under hypotheses (2.6.1), (2.6.3) for $\varrho = k+\varkappa$ and (3.3.15), $A \in L(W^k_{p,A}(0,\infty), W^k_p(0,\infty))$ is a Fredholm operator with index

$$\operatorname{ind} A = \xi + \xi_\infty + \xi(p,k+\varkappa) + \xi_\infty(p) - \min(k,-\varkappa)-1 ,$$

and it is not normally solvable if one of these conditions is violated.

Furthermore, as a consequence of Theorems 2.4.6, 3.3.1 and 3.1.1 (with $X^{(\infty)} = Y^{(\infty)} = S[0,\infty))$ one obtains

<u>Theorem 3.4.2.</u> $A \in L(S[0,\infty))$ is a Fredholm operator with index $\xi + \xi_\infty + \varkappa - 1$.

<u>3.4.2.</u> We now consider some examples.

<u>Example 3.4.3.</u> $A = \sum_{\substack{n+i \leq l \\ \sigma + n\delta + i \in \mathbb{N}_0}} a_{in} x^{\sigma + n\delta + i} D^i$,

where $\sigma \in \mathbb{Z}$, $i,n \in \mathbb{N}_0$, $\delta > 0$, $a_{in} \in \mathbb{C}$ and $a_{l0} \neq 0$.
We first study the operator A on an interval (0,b), b> 0. With the notation of Chap. 2, $l_0 = l$ and $\varkappa = -\sigma$. The characteristic equation (2.1.12) takes the form

$$\sum_{0 \leq i \leq l,\ \sigma+i \in \mathbb{N}_0} a_{i0}\, \mu(\mu-1)\dots(\mu-i+1) = 0 . \qquad (3.4.2)$$

Furthermore,

$$W^k_{p,A}(0,b) = \{ y \in \mathcal{D}'(0,b) : x^{\sigma+i} D^{i+j} y \in L_p(0,b), i+\sigma \in \mathbb{N}_0 ,$$

$i=0,\dots,l,\ j=0,\dots,k\}$.

If the roots μ_i of (3.4.2) satisfy the condition

$$\mathrm{Re}\,\mu_i \neq -1/p+k-\sigma\ ,\ i=1,\dots,l\ , \qquad (3.4.3)$$

then Theorem 2.6.6 implies that $A\in L(W^k_{p,A}(0,b),W^k_p(0,b))$ is a Fredholm operator with index $\zeta(p,k-\sigma)-\min(k,\sigma)$, where $\zeta(p,k-\sigma)$ denotes the number of roots satisfying $\mathrm{Re}\,\mu_i > -1/p+k-\sigma$. This assertion is also true when $\delta=0$ and $a_{in}=0$ for $n>0$.

We now consider A on the interval (b,∞). Assume that the roots of the equation

$$\sum_{n+i=l,\ \sigma+n\delta\in\mathbb{N}_0} a_{in}(-\gamma)^i = 0 \qquad (3.4.4)$$

satisfy the condition

$$\mathrm{Re}\,\gamma_i \neq 0\ ,\ i=1,\dots,l\ . \qquad (3.4.5)$$

Then $a_{ol}\neq 0$, and with the notation of 3.3.1, we have $l_\infty=0$, $\varkappa_\infty = -\sigma-l\delta$ and equations (3.3.13) reduce to (3.4.4). Moreover, $q_i = 1-\delta$ $(i=1,\dots,l)$, where q_i are the numbers defined by (3.3.11). Hence

$$W^k_{p,A}(b,\infty) = \{\, y\in\mathcal{D}'(b.\infty)\ :\ x^{\sigma+(l-i)\delta+i}D^{i+j}y\in L_p(b,\infty)\ ,$$

$$i=0,\dots,l,\ j=0,\dots,k\,\}\ .$$

Since $\sigma+l\in\mathbb{N}_0$ by $a_{lo}\neq 0$ and $\sigma+l\delta\in\mathbb{N}_0$ by $a_{ol}\neq 0$, $\sigma+(l-i)\delta+i\geq 0$ for $i=0,\dots,l$. Thus Remark 3.3.4 implies that we may assume $\sigma+(l-i)\delta+i\in\mathbb{N}_0$ in the definition of $W^k_{p,A}(b,\infty)$. To summarize,

$$W^k_{p,A}(0,\infty) = \{\, y\in\mathcal{D}'(0,\infty)\ :\ x^{\sigma+n\delta+i}D^{i+j}y\in L_p(0,\infty)\ ,$$

$$\sigma+n\delta+i\in\mathbb{N}_0,\ 0\leq n+i\leq l,\ j=0,\dots,k\,\}\ .$$

If ζ_∞ denotes the number of roots of (3.4.4) satisfying $\mathrm{Re}\,\gamma_i>0$, then Theorem 3.4.1 implies the following result:

Under the assumptions (3.4.3) and (3.4.5), $A\in L(W^k_{p,A}(0,\infty),W^k_p(0,\infty))$ is a Fredholm operator with index $\zeta(p,k-\sigma)+\zeta_\infty-\min(k,\sigma)-l$.

<u>Example 3.4.4.</u> $A=\sum_{0\leq i\leq l} a_i x^i D^i\ ,\ a_i\in\mathbb{C}\ ,\ a_l\neq 0$.

Then $l_\infty = 1$ and $\varkappa_\infty = 0$. Equation (3.3.12) takes the form

$$\sum_{0 \le i \le l} a_i(-\gamma)(-\gamma-1)\dots(-\gamma-i+1) = 0 \tag{3.4.6}$$

and one gets the characteristic equation of A at the origin by replacing $-\gamma$ by γ in (3.4.6). Moreover,

$$W^k_{p,A}(0,\infty) = \{ y \in W^k_p(0,\infty) \; : \; x^i D^i y \in W^k_p(0,\infty), \; i=1,\dots,l \} .$$

Applying the result of Example 3.4.3 (for $\sigma = \delta = 0$ and the interval (0,b)) and Theorem 3.4.1, we obtain:

$A \in L(W^k_{p,A}(0,\infty), W^k_p(0,\infty))$ is normally solvable if and only if the roots of (3.4.6) satisfy $\operatorname{Re}\gamma_i \neq 1/p, 1/p-k$ (i=1,...,l). Under this condition, A has index $-z(p,k)$, where $z(p,k)$ denotes the number of γ_i's satisfying $1/p-k < \operatorname{Re}\gamma_i < 1/p$.

<u>Example 3.4.5.</u> $A = \sum_{0 \le i \le l, iq \in \mathbb{N}_0} a_i x^{iq} D^i$, $q > 1$, $a_i \in \mathbb{C}$, $a_l \neq 0$.

We assume that the roots γ_i of the equation

$$\sum_{0 \le i \le l, iq \in \mathbb{N}_0} a_i \gamma^i = 0 \tag{3.4.7}$$

satisfy the condition

$$\operatorname{Re}\gamma_i \neq 0, \; i=1,\dots,l . \tag{3.4.8}$$

Then $a_0 \neq 0$, $l_0 = \varkappa = 0$, and in view of Remark 2.6.9

$$W^k_{p,A}(0,b) = \{ y \in W^k_p(0,b) \; : \; x^{iq} D^{i+j} y \in L_p(0,b), iq \in \mathbb{N}_0, i=1,\dots,l, \\ j=0,\dots,k \} .$$

Let ξ be the number of γ_i's satisfying $\operatorname{Re}\gamma_i > 0$. By Theorem 2.6.6, $A \in L(W^k_{p,A}(0,b), W^k_p(0,b))$ is a Fredholm operator with index ξ if condition (3.4.8) is satisfied.

Furthermore, $l_\infty = l$, $\varkappa_\infty = -l(q-1)$ and

$$W^k_{p,A}(b,\infty) = \{ y \in \mathcal{D}'(b,\infty) \; : \; x^{i-\varkappa_\infty} D^{i+j} y \in L_p(b,\infty), \\ i=0,\dots,l, \; j=0,\dots,k \} .$$

Using Hardy's inequality (see Stein [1, Appendix A.4]), one can show that

$$W^k_{p,A}(b,\infty) = \{ y \in \mathcal{D}'(b,\infty) : x^{iq}D^{i+j}y \in L_p(b,\infty),\ iq \in \mathbb{N}_0,$$
$$i=0,\dots,l,\ j=0,\dots,k \}$$

including the equivalence of the canonical norms. The characteristic equation of A at ∞ is $-\gamma(-\gamma-1)\dots(-\gamma-l+1) = 0$ and their roots γ_i obviously satisfy $\mathrm{Re}\,\gamma_i < 1/p - \varkappa_\infty$ $(i=1,\dots,l)$. Hence $\dot{A} \in L(W^k_{p,A}(b,\infty), W^k_p(b,\infty))$ is an isomorphism by virtue of Theorem 3.3.3. To summarize,

$$W^k_{p,A}(0,\infty) = \{ y \in W^k_p(0,\infty) : x^{iq}D^i y \in W^k_p(0,\infty),\ iq \in \mathbb{N}_0, i=1,\dots,l \}$$

and Theorem 3.4.1 yields the following result:

Under hypothesis (3.4.8), $A \in L(W^k_{p,A}(0,\infty), W^k_p(0,\infty))$ is an injective Fredholm operator with deficiency $l-\xi$.

3.5. Differential operators with a finite number of singularities on $(-\infty,\infty)$

Let x_j $(j=1,\dots,n)$ be distinct points on the real axis, $q_j \in \mathbb{N}$ and

$$\beta(x) = (x-x_1)^{q_1}\dots(x-x_n)^{q_n} .$$

For a function $y \in \bar{C}^\infty(\mathbb{R})$ with the asymptotic expansion (3.3.2) as $x \to -\infty$, set $v_{-\infty}(y) = M$ when $d_0 \neq 0$ and $v_{-\infty}(y) = -\infty$ when $d_j = 0$ for all $j \in \mathbb{N}_0$. We consider differential operators of the form

$$A = \sum_{0 \le i \le l} a_i(x)D^i,\ a_i \in \bar{C}^\infty(\mathbb{R})\ (i=0,\dots,l),\ a_l = \beta \bar{a}_l ,$$
$$\bar{a}_l(x) \neq 0\ (x \in \mathbb{R}),\ v_\infty(a_l) > -\infty ,\ v_{-\infty}(a_l) > -\infty . \tag{3.5.1}$$

With A we associate the formal differential operators

$$\hat{A}_j = \sum_{0 \le i \le l} \hat{a}_{ji}D^i ,\ \hat{a}_{ji} = \sum_{k \ge 0} a_i^{(k)}(x_j)(x-x_j)^k/k! .$$

Substituting $x \to x+x_j$ in $\hat{A}_j$, we obtain an operator of the form (2.3.1) and denote its characteristic factors by $\tilde{\lambda}_i^{(j)}$ $(i=1,\dots,l)$. The functions $\tilde{\lambda}_{ji}(x) = \tilde{\lambda}_i^{(j)}(x-x_j)$ $(i=1,\dots,l)$ are called the characteristic factors of A at the singular point x_j. Set further

$$\varkappa_j = \max_{0 \le i \le l}(i-v_j(a_i)),\ j=1,\dots,n ,$$

where $v_j(a)$ denotes the order of the zero of a at x_j. Let $\xi_j^{\pm}$ be the number of characteristic factors at x_j satisfying

Re $\tilde{\lambda}_{ji}(x) \to +\infty$ as $x \to x_j \pm 0$, and set $\xi_j = \xi_j^+ + \xi_j^-$.

Furthermore, with (3.5.1) we associate the formal operator

$$\hat{A}_{-\infty} = \sum_{0 \le i \le l} \hat{a}_{-\infty i} D^i ,$$

where $\hat{a}_{-\infty i}$ denote the asymptotic expansions of a_i as $x \to -\infty$. Substituting $x \to 1/x$ in $\hat{A}_{-\infty}$, one obtains a differential polynomial of the form (2.1.2) with coefficients in $\hat{\alpha}_1$, the characteristic factors of which will be denoted by $\tilde{\lambda}_i^-$. Then the functions $\tilde{\lambda}_{-\infty i}(x) = \tilde{\lambda}_i^-(1/x)$ $(i=1,\dots,l)$ are called the characteristic factors of A at $-\infty$. Let $\xi_{-\infty}$ be the number of characteristic factors at $-\infty$ satisfying Re $\tilde{\lambda}_{-\infty i}(x) \to +\infty$ as $x \to -\infty$. For the operator (3.5.1), we now have the following result.

<u>Theorem 3.5.1.</u> $A \in L(S(\mathbb{R}))$ is a Fredholm operator with index

$$\operatorname{ind} A = \sum_{1 \le j \le n} (\xi_j + \varkappa_j) + \xi_\infty + \xi_{-\infty} - (n+1)l . \qquad (3.5.2)$$

<u>Proof.</u> In analogy to Theorem 2.4.6 one obtains that $A \in L(C^\infty(U_j))$ is a Fredholm operator with index $\xi_j + \varkappa_j$, where U_j is a closed interval containing x_j as an interior point such that $x_k \notin U_j$ for all $k \ne j$. Furthermore, as in Theorem 3.3.1 one can show that $A \in L(S(-\infty,b])$ is a Fredholm operator with index $\xi_{-\infty}$ when $b < x_j$ for all j. Finally, applying Theorems 3.3.1 and 3.1.1, we conclude formula (3.5.2). □

<u>Corollary 3.5.2.</u> Let $p_i(x)$ $(i=0,\dots,l)$ be polynomials in x and $p_l \not\equiv 0$. Then $p_l(x)D^l + \dots + p_1(x)D + p_0(x)$ is always a Fredholm operator in $S(\mathbb{R})$.

<u>3.6. Comments and references</u>

<u>3.1.</u> In certain special cases the local principle of Theorem 3.1.1 was used earlier in Višik and Grušin [1] and Bolley, Camus and Helffer [1]. A similar method has been applied to the computation of deficiency indices of symmetric ordinary differential operators; see e.g. Neumark [1, p. 206].

3.2. Here we followed Elschner [3]. The special case of generalized Legendre operators was considered by Triebel [1], [2]; see also the presentation in his monograph [4]. The essential spectrum of the operators $\mathcal{A}$ for $l_1 = 2m$, $l_2 = 0$ and real coefficients c_i was studied by Müller-Pfeiffer [1], using Hilbert space methods. By other methods, the first part of Theorem 3.2.7 was proved in Triebel [1]. The second assertion of this theorem is also valid for $1 < p < 2$ (cf. Elschner [3]) and answers a question contained in Triebel [2]. For further results on selfadjointness and domains of definition of the operators $\mathcal{A}^j$, we refer to Triebel [1], [4] for generalized Legendre operators and to Elschner [3] for the general case.

3.3 - 5. Here we essentially followed Elschner [5]. The results of Example 3.4.3 for $p = 2$ are contained in Bolley, Camus and Helffer [1], and in certain special cases they can be found in Bolley and Camus [1]. The results of Example 3.4.5 for $k=0$ and $p=2$ were proved in Bolley, Camus and Helffer [2].

4. ELLIPTIC DIFFERENTIAL OPERATORS IN $\mathbb{R}^n$ DEGENERATING AT ONE POINT

This chapter is devoted to some examples of elliptic partial differential operators which degenerate at the origin. Our first example is

$$A = r^2\Delta + ar\,\partial/\partial r + b, \quad a,b \in \mathbb{C} \quad , \tag{4.0.1}$$

where $r = |x| = (x_1^2 + \ldots + x_n^2)^{1/2}$, $x = (x_1,\ldots,x_n) \in \mathbb{R}^n$, $n \geq 2$ and $\Delta = \sum_i \partial^2/\partial x_i^2$. In order to study the solvability properties of (4.0.1) in the spaces L_2 and C^∞, we use spherical harmonics expansions and associate an infinite set of second order Euler operators with A. In contrast to the results of Chapters 1 and 2 for ordinary differential operators with one singularity, the operator (4.0.1) acting in the space of C^∞ functions which are flat at the origin is not surjective and even not normally solvable. Furthermore, we consider a class of degenerate elliptic operators of higher order which is a natural generalization of the (one-dimensional) Euler operators and contains (4.0.1) as a special case. Using the Mellin transform and a result of Agranovič and Višik on elliptic differential operators depending on a parameter, we investigate the Fredholm property in Sobolev spaces. Finally, we give an example of an elliptic differential operator in $\mathbb{R}^2$ whose symbol vanishes to finite order at the origin, which is not locally solvable there. Note that in view of Theorem 2.5.1 such an example cannot exist in the one-dimensional case.

4.1. Preliminaries on function spaces and the Mellin transform

4.1.1. We first introduce certain n-dimensional analogues of the function spaces defined in 1.1. Henceforth by Ω we denote the whole space $\mathbb{R}^n$ $(n \geq 2)$ or a bounded domain in $\mathbb{R}^n$ such that its boundary $\partial\Omega$ is an (n-1)-dimensional infinitely differentiable manifold and Ω is locally on one side of $\partial\Omega$. For an n-tuple of non-negative integers $\alpha = (\alpha_1,\ldots,\alpha_n)$, we set

$$D^\alpha = D_x^\alpha = (\partial/\partial x_1)^{\alpha_1}\ldots(\partial/\partial x_n)^{\alpha_n}, \quad |\alpha| = \alpha_1+\ldots+\alpha_n ,$$

$$x^\alpha = x^{\alpha_1}\ldots x^{\alpha_n} .$$

For any real number $k \geq 0$, let $W_2^k(\Omega)$ be the usual Sobolev space, i.e. the completion of $C_0^\infty(\mathbb{R}^n)$ (resp. $C^\infty(\bar{\Omega})$ when Ω is bounded) with respect to the norm

$$\|u\|_{k,\Omega} = \left\{ \sum_{|\alpha|\leq m} \int_\Omega |D^\alpha u|^2 dx + \sum_{|\alpha|=m} \int_{\Omega\times\Omega} \frac{|D^\alpha u(x) - D^\alpha u(y)|^2}{|x-y|^{n+2\theta}} dxdy \right\}^{\frac{1}{2}} \tag{4.1.1}$$

where $k = m+\theta$, $m \in \mathbb{N}_0$, $\theta \in [0,1)$ and the second term on the right-hand side of (4.1.1) is omitted for $\theta = 0$. We set $W_2^0 = L_2$. Furthermore, for $l \in \mathbb{N}$, we introduce the weighted Sobolev space

$$W_{2,l}^k(\Omega) = \{ u \in W_2^k(\Omega) : x^\beta D^\alpha u \in W_2^k(\Omega), \quad |\alpha| = |\beta| \leq l \}$$

with the canonical norm

$$\|u\|_{k,l,\Omega} = \sum_{|\alpha|=|\beta|\leq l} \|x^\beta D^\alpha u\|_{k,\Omega} . \tag{4.1.2}$$

We set $W_{2,l}^0 = L_{2,l}$. Note that the terms $x^\beta D^\alpha u$ are defined in the sense of distributions. When Ω is bounded and $0 \bar{\in} \bar{\Omega}$, the spaces $W_{2,l}^k(\Omega)$ and $W_2^{k+l}(\Omega)$ coincide (algebraically and topologically).

Let now $0 \in \Omega$. By Sobolev's theorem,

$$\mathring{W}_2^k(\Omega) = \{ u \in W_2^k(\Omega) : D^\alpha u(0) = 0, \quad |\alpha| < k - n/2 \}$$

is a closed subspace of $W_2^k(\Omega)$ of finite codimension. Furthermore, we introduce the closed subspace

$$\mathring{W}_{2,l}^k(\Omega) = \{ u \in W_{2,l}^k(\Omega) : D^\alpha u(0) = 0, \quad |\alpha| < k - n/2 \}$$

of $W_{2,l}^k(\Omega)$. Finally, let

$$\mathring{C}^\infty(\Omega) = \{ u \in C^\infty(\bar{\Omega}) : 0 \bar{\in} \operatorname{supp} u \}$$

when Ω is bounded and

$$\mathring{C}^\infty(\mathbb{R}^n) = \{ u \in C_0^\infty(\mathbb{R}^n) : 0 \bar{\in} \operatorname{supp} u \} .$$

<u>Lemma 4.1.1.</u> $\mathring{C}^\infty(\Omega)$ is dense in $\mathring{W}_2^k(\Omega)$ for any $k \geq 0$.

<u>Proof.</u> Let first $\Omega = \mathbb{R}^n$. Consider an arbitrary continuous linear

functional F on $\overset{\circ}{W}{}_2^k$ and let $\tilde{F}$ be a continuous extension of F on W_2^k. Assume that $F(u) = 0$ for all $u \in \overset{\circ}{C}{}^\infty$. We have to show that $F(u) = 0$ for any $u \in \overset{\circ}{W}{}_2^k$. Since $\tilde{F}(u) = 0$ for all $u \in \overset{\circ}{C}{}^\infty$, the support of the distribution $\tilde{F}$ consists only of the point 0. Hence $\tilde{F}$ is a linear combination of derivatives of the Dirac measure at the origin of order $\leq N$ for some $N \in \mathbb{N}_0$. Since $\tilde{F}$ is continuous on W_2^k, $N < k-n/2$. Consequently, $F(u) = \tilde{F}(u) = 0$ for all $u \in \overset{\circ}{W}{}_2^k$ Q.E.D.

Let Ω be a bounded domain and $u \in \overset{\circ}{W}{}_2^k(\Omega)$. Then one can extend u to a function $\tilde{u} \in \overset{\circ}{W}{}_2^k(\mathbb{R}^n)$, and choosing a sequence $\{u_m\} \subset \overset{\circ}{C}{}^\infty(\mathbb{R}^n)$ which converges to $\tilde{u}$ in $\overset{\circ}{W}{}_2^k(\mathbb{R}^n)$, one obtains that $\{u_m|_{\bar{\Omega}}\} \subset \overset{\circ}{C}{}^\infty(\Omega)$ approximates u in $\overset{\circ}{W}{}_2^k(\Omega)$. □

<u>Lemma 4.1.2.</u> $\overset{\circ}{C}{}^\infty(\Omega)$ is dense in $L_{2,1}(\Omega)$.

<u>Proof.</u> If $\Omega = \mathbb{R}^n$, we choose a sequence $\{\psi_m(r)\} \subset C_0^\infty(\mathbb{R}^n)$ such that $\psi_m(r) = 1$ for $r = |x| \leq m$ and $|x^\beta D^\alpha \psi_m(r)| \leq c$ for all $x \in \mathbb{R}^n$, $|\alpha| = |\beta| \leq 1$ and $m \in \mathbb{N}$. Then it is easy to check that $\|u - \psi_m u\|_{0,1,\mathbb{R}^n} \to 0$ as $m \to \infty$ for any $u \in L_{2,1}(\mathbb{R}^n)$. Thus it is sufficient to prove the lemma for a bounded domain Ω.

For $u \in L_{2,1}(\Omega)$, we define the sequence $u_m = \chi(mr)u$ $(m \in \mathbb{N})$ in $\overset{\circ}{C}{}^\infty$, where $\chi(r) \in C^\infty(\mathbb{R}^n)$ is such that $0 \leq \chi \leq 1$, $\chi(r) = 1$ for $r \geq 2$ and $\chi(r) = 0$ for $r \leq 1$. Similarly as in the proof of Lemma 3.2.3 one can verify that

$$\|x^\beta D^\alpha (u - u_m)\|_{0,\Omega} \to 0 \quad (m \to \infty), \quad |\alpha| = |\beta| \leq 1 . \; \square$$

<u>4.1.2.</u> Let S^{n-1} be the unit sphere in $\mathbb{R}^n$. We introduce the spherical coordinates of $x \in \mathbb{R}^n$ by setting $x = r\theta$ with $r = |x|$ and $\theta = (\theta_1, \ldots, \theta_n) \in S^{n-1}$, where

$$\begin{aligned} \theta_1 &= \cos \vartheta_1 \\ \theta_2 &= \sin \vartheta_1 \cos \vartheta_2 \\ &\ldots \qquad \ldots \\ \theta_{n-1} &= \sin \vartheta_1 \ldots \sin \vartheta_{n-2} \cos \vartheta_{n-1} \\ \theta_n &= \sin \vartheta_1 \ldots \sin \vartheta_{n-2} \sin \vartheta_{n-1} , \end{aligned} \tag{4.1.3}$$

where $0 \leq \vartheta_j \leq \pi$ $(j=1,\ldots,n-2)$, $0 \leq \vartheta_{n-1} < 2\pi$ and $n \geq 4$. It is clear how to modify (4.1.3) for $n = 2$ and $n = 3$. We have

$$\begin{pmatrix} \partial/\partial x_1 \\ \cdot \\ \cdot \\ \cdot \\ \partial/\partial x_n \end{pmatrix} = r^{-1} J \begin{pmatrix} r\,\partial/\partial r \\ \partial/\partial \vartheta_1 \\ \cdot \\ \cdot \\ \partial/\partial \vartheta_{n-1} \end{pmatrix} \tag{4.1.4}$$

where each element of the n-by-n matrix J belongs to $C^\infty(S^{n-1})$.

The Laplacian $\Delta = \sum_i \partial^2/\partial x_i^2$ in $\mathbb{R}^n$ can be written in spherical coordinates

$$\Delta = \partial^2/\partial r^2 + (n-1)r^{-1}\partial/\partial r - r^{-2}\delta\,, \tag{4.1.5}$$

where

$$\delta = -\sum_{1 \le i < n} q_i^{-1} \sin^{1-n+i}\vartheta_i \;\partial/\partial\vartheta_i \;\sin^{n-1-i}\vartheta_i \;\partial/\partial\vartheta_i\,,$$

$$q_1 = 1,\; q_i = (\sin\vartheta_1 \ldots \sin\vartheta_{i-1})^2, \quad i \ge 2\,,$$

is the Beltrami operator on S^{n-1}. Let $L_2(S^{n-1})$ be the Hilbert space of square integrable functions on S^{n-1} with norm

$$\|u\|_{0,S^{n-1}} = (u,u)^{1/2}$$

and scalar product

$$(u,v) = \int_{S^{n-1}} u\bar{v}\,d\theta = \int_{S^{n-1}} u\bar{v}\,dS^{n-1}\,.$$

The closure of the operator

$$1+\delta\,,\; D(1+\delta) = C^\infty(S^{n-1})$$

in $L_2(S^{n-1})$ is selfadjoint and positive definite (see Triebel [3, § 31]) and will also be denoted by $1+\delta$. For $k \in \mathbb{N}_0$, we define the Sobolev space $W_2^k(S^{n-1})$ to be the domain of definition of the operator $(1+\delta)^{k/2}$ equipped with the norm

$$\|u\|_{k,S^{n-1}} = \|(1+\delta)^{k/2}u\|_{0,S^{n-1}}\;; \tag{4.1.6}$$

see Lions and Magenes [1, Chap. 1.7.3].

Let Ω_R be the open ball $\{x \in \mathbb{R}^n : |x| < R\}$, $0 < R \le \infty$. Then the scalar product in $L_2(\Omega_R)$ is given by

$$(u,v)_{\Omega_R} = \int_{\Omega_R} u\bar{v}dx = \int_0^R \int_{S^{n-1}} u\bar{v}r^{n-1}d\theta\, dr \,. \qquad (4.1.7)$$

It follows from Lemma 4.1.2 and the next lemma that, for $u \in L_{2,1}(\Omega_R)$ and $h + j \leq 1$, $(1+\delta)^{h/2}(r\partial/\partial r)^j u$ is defined as a distribution on Ω_R and that

$$|u|_{0,1,\Omega_R} = \sum_{h+j\leq 1} \|(1+\delta)^{h/2}(r\partial/\partial r)^j u\|_{0,\Omega_R}$$

is equivalent to the norm (4.1.2) in $L_{2,1}(\Omega_R)$. Henceforth c and c_1 denote various constants independent of u.

<u>Lemma 4.1.3.</u> For all $u \in \mathring{C}^\infty(\Omega_R)$ the following inequalities hold:

$$c_1|u|_{0,1,\Omega_R} \leq \|u\|_{0,1,\Omega_R} \leq c|u|_{0,1,\Omega_R} \,. \qquad (4.1.8)$$

<u>Proof.</u> Let first $\Omega_R = \mathbb{R}^n$ and $u \in \mathring{C}^\infty(\mathbb{R}^n)$. (4.1.5) implies

$$\begin{aligned} \|(1+\delta)^{1/2}u\|^2_{0,\mathbb{R}^n} &= ((1+\delta)u,u)_{\mathbb{R}^n} = \|u\|^2_{0,\mathbb{R}^n} \\ &+ (n-1)(r\partial u/\partial r,u)_{\mathbb{R}^n} - (r^2\Delta u,u)_{\mathbb{R}^n} + (r^2\partial^2u/\partial r^2,u)_{\mathbb{R}^n} \,. \end{aligned} \qquad (4.1.9)$$

The first two terms on the right-hand side of (4.1.9) are less than $c\|u\|^2_{0,1,\mathbb{R}^n}$. Integrating by parts, we see that the last two terms can be estimated in the same way. Since $r\partial/\partial r = \sum_i x_i\partial/\partial x_i$, we obviously have $\|r\partial u/\partial r\|_{0,\mathbb{R}^n} \leq \|u\|_{0,1,\mathbb{R}^n}$ so that

$$|u|_{0,1,\mathbb{R}^n} \leq c\|u\|_{0,1,\mathbb{R}^n} \,.$$

To prove the second inequality in (4.1.8) for l = 1, we introduce the operators

$$\partial_{x_i} = r\partial/\partial x_i - x_i\partial/\partial r \,, \quad i=1,\ldots,n \,.$$

Using the representation

$$x_j\partial/\partial x_i = r^{-1}x_j\partial_{x_i} + r^{-2}x_ix_jr\partial/\partial r \,,$$

we obtain the estimate

$$\|u\|^2_{0,1,\mathbb{R}^n} \leq c\{\|u\|^2_{0,\mathbb{R}^n} + \|r\partial u/\partial r\|^2_{0,\mathbb{R}^n} + \textstyle\sum_i \|\partial_{x_i}u\|^2_{0,\mathbb{R}^n}\}. \qquad (4.1.10)$$

Furthermore, it is easy to check that

$$\sum_i (\partial_{x_i})^2 = -\delta\,,\quad (u, \partial_{x_i} u)_{\mathbb{R}^n} = -(\partial_{x_i} u, u)_{\mathbb{R}^n} + (a_i(\theta)u, u)_{\mathbb{R}^n}$$

with some $a_i(\theta) \in C^\infty(S^{n-1})$. Therefore

$$\|u\|^2_{0,\mathbb{R}^n} + \sum_i \|\partial_{x_i} u\|^2_{0,\mathbb{R}^n} \leq c\{\|u\|^2_{0,\mathbb{R}^n} + \|(1+\delta)^{1/2}u\|^2_{0,\mathbb{R}^n}$$

$$+ \sum_i \|u\|_{0,\mathbb{R}^n} \|\partial_{x_i} u\|_{0,\mathbb{R}^n}\}$$

which implies

$$\|u\|^2_{0,\mathbb{R}^n} + \sum_i \|\partial_{x_i} u\|^2_{0,\mathbb{R}^n} \leq c_1\{\|u\|^2_{0,\mathbb{R}^n} + \|(1+\delta)^{1/2}u\|^2_{0,\mathbb{R}^n}\}.$$

Together with (4.1.10), this gives

$$\|u\|_{0,1,\mathbb{R}^n} \leq c|u|_{0,1,\mathbb{R}^n}.$$

Repeating the above arguments, we obtain (4.1.8) by induction on l.

Finally, let $R < \infty$ and $u \in \overset{\circ}{C}{}^\infty(\Omega_R)$. Choose a function $\chi(r) \in C^\infty(\mathbb{R}^n)$ such that $\chi(r) = 1$ for $r \leq R/3$ and $\chi(r) = 0$ for $r \geq R/2$, and let $\Omega' = \{x \in \mathbb{R}^n : R/3 < r < R\}$. Then it is easy to check that

$$\|\chi u\|_{0,1,\Omega_{R/2}} + \|(1-\chi)u\|_{1,\Omega'}$$

is an equivalent norm in $L_{2,1}(\Omega_R)$, and by Theorem 1.3 in Lions and Magenes [1, Chap. 1],

$$c_1\|(1-\chi)u\|_{1,\Omega'} \leq \sum_{h+j\leq 1} \|(1+\delta)^{h/2}(r\partial/\partial r)^j(1-\chi)u\|_{0,\Omega'}$$
$$\leq c\|(1-\chi)u\|_{1,\Omega'}.$$

Since further

$$c_1|\chi u|_{0,1,\Omega_{R/2}} \leq \|\chi u\|_{0,1,\Omega_{R/2}} \leq c|\chi u|_{0,1,\Omega_{R/2}}$$

in view of $\chi u \in \overset{\circ}{C}{}^\infty(\mathbb{R}^n)$ and $\operatorname{supp} \chi u \subset \overline{\Omega}_{R/2}$, we obtain (4.1.8) for $R < \infty$. □

4.1.3. For an open set $\Omega \subset \mathbb{R}^n$ and $\rho, k \in \mathbb{R}$, define the weighted L_2 space

$$L_2^{\varrho,k}(\Omega) = \{r^{\varrho}(1+r)^{k-\varrho}u : u \in L_2(\Omega)\}$$

with norm $\| r^{-\varrho}(1+r)^{\varrho-k}u \|_{0,\Omega}$. Furthermore, for $\varrho \geq 0$ and $l \in \mathbb{N}$, we define the Banach space

$$L_{2,l}^{\varrho,k}(\Omega) = \{u \in L_2^{\varrho,k}(\Omega) : x^{\beta}D^{\alpha}u \in L_2^{\varrho,k}(\Omega),\ |\alpha| = |\beta| \leq l\}$$

with the canonical norm

$$\| u \|_{L_{2,l}^{\varrho,k}} = \sum_{|\alpha| = |\beta| \leq l} \| r^{-\varrho}(1+r)^{\varrho-k}x^{\beta}D^{\alpha}u \|_{0,\Omega} .$$

We set $L_2^{\varrho,\varrho} = L_2^{\varrho}$ and $L_{2,l}^{\varrho,\varrho} = L_{2,l}^{\varrho}$.

Remark 4.1.4. Similarly as in Lemma 4.1.2 one can prove that $\mathring{C}^{\infty}(\mathbb{R}^n)$ is also dense in $L_{2,l}^{\varrho,k}(\mathbb{R}^n)$. Furthermore, replacing the L_2 norms by the norms in $L_2^{\varrho,k}$, one obtains an extension of Lemma 4.1.3 to the spaces $L_{2,l}^{\varrho,k}(\mathbb{R}^n)$.

For $u \in C_0^{\infty}(\mathbb{R}^n)$, we define the Fourier transform by

$$(\mathcal{F}u)(\xi) = (2\pi)^{-n/2} \int_{\mathbb{R}^n} e^{-ix\xi}u(x)dx ,$$

where $x\xi = \sum_j x_j\xi_j$, $\xi = (\xi_1,\dots,\xi_n)$ and $i = \sqrt{-1}$. Then the relations

$$\mathcal{F}(x^{\beta}D_x^{\alpha}u) = i^{|\alpha|+|\beta|}D_{\xi}^{\beta}\xi^{\alpha}\mathcal{F}u ,\quad u \in C_0^{\infty}(\mathbb{R}^n) \tag{4.1.11}$$

hold. The Fourier transform extends to an isometric isomorphism of $L_2(\mathbb{R}^n)$ onto itself and to an isomorphism of $W_2^k(\mathbb{R}^n)$ onto $L_2^{0,-k}(\mathbb{R}^n)$ (cf. e.g. Triebel [3, § 10]). Since $C_0^{\infty}(\mathbb{R}^n)$ is dense in $L_{2,1}(\mathbb{R}^n)$, (4.1.11) is valid for any $u \in L_{2,1}(\mathbb{R}^n)$ and $|\alpha| = |\beta| \leq 1$. Consequently, $\mathcal{F}$ is an isomorphism of $W_{2,1}^k(\mathbb{R}^n)$ onto $L_{2,1}^{0,-k}(\mathbb{R}^n)$ since $D^{\beta}x^{\alpha}$, $|\alpha| = |\beta|$, is a linear combination of the terms $x^{\tilde{\alpha}}D^{\tilde{\beta}}$, $|\tilde{\alpha}| = |\tilde{\beta}| \leq |\alpha|$.

4.1.4. For $u \in C_0^{\infty}(0,\infty)$, we define the Mellin transform by

$$(\mathcal{M}u)(z) = (2\pi)^{-1/2} \int_0^{\infty} r^{-iz-1}u(r)dr$$

which is an entire function of $z \in \mathbb{C}$. Note that

$$\mathcal{M} = \mathcal{F} \circ \Phi \quad , \quad \Phi : u(r) \to u(e^{\bar{r}}),\ \bar{r} = \ln r ,$$

so the properties of the Mellin transform follow from the corresponding ones of the Fourier transform. Concerning the inverse of the Mellin transform, we have

$$u(r) = (2\pi)^{-1/2} \int_{-\infty}^{\infty} r^{iz}(\mathcal{M}u)(z)dt, \quad z = t + i\tau \tag{4.1.12}$$

for any $\tau \in \mathbb{R}$. Moreover, for any $u \in C_0^\infty(0,\infty)$, the relations

$$\int_{-\infty}^{\infty} |\mathcal{M}u(z)|^2 dt = \int_0^\infty r^{2\tau-1} |u(r)|^2 dr, \quad z = t + i\tau, \tag{4.1.13}$$

$$\mathcal{M}(rdu/dr) = iz\mathcal{M}u \tag{4.1.14}$$

hold (see Eskin [1, § 2]).

$L_2(\{\mathrm{Im} z = \varrho\})$ will denote the space of square integrable functions on the line $\mathrm{Im}\, z = \varrho$. Let

$$\Pi_{\sigma,\varrho} = \{z \in \mathbb{C} : \sigma < \mathrm{Im}\, z < \varrho\}, \quad \sigma < \varrho,$$

and let $A_2(\Pi_{\sigma,\varrho})$ be the Banach space of all analytic functions u in the strip $\Pi_{\sigma,\varrho}$ for which the norm

$$\|u\|_{A_2(\Pi_{\sigma,\varrho})} = \sup_{\sigma<\tau<\varrho} \left(\int_{-\infty}^{\infty} |u(z)|^2 dt\right)^{1/2}, \quad z = t + i\tau,$$

is finite. For any $u \in A_2(\Pi_{\sigma,\varrho})$, the boundary values

$$u(t+i\sigma) \in L_2(\{\mathrm{Im}\, z = \sigma\}), \quad u(t+i\varrho) \in L_2(\{\mathrm{Im}\, z = \varrho\})$$

exist in the sense that (see Hoffman [1])

$$\lim_{\tau\to\sigma+} \int_{-\infty}^{\infty} |u(t+i\tau) - u(t+i\sigma)|^2 dt = 0,$$

$$\lim_{\tau\to\varrho-} \int_{-\infty}^{\infty} |u(t+i\tau) - u(t+i\varrho)|^2 dt = 0.$$

<u>Proposition 4.1.5.</u> (i) $\mathcal{M}$ extends to an isometric isomorphism of $L_2^\varrho(0,\infty)$ onto $L_2(\{\mathrm{Im}\, z = -\varrho + 1/2\})$.

(ii) For $\sigma < \varrho$, $\mathcal{M}$ is an isomorphic map of $L_2^{\varrho,\sigma}(0,\infty)$ onto $A_2(\Pi_{1/2-\varrho,1/2-\sigma})$, and formulas (4.1.12), (4.1.13) hold for all $u \in L_2^{\varrho,\sigma}(0,\infty)$ and $\tau \in [1/2-\varrho, 1/2-\sigma]$.

This follows from Theorems 9c and 9d in Widder [1, Chap. VI].

Let $u(r\theta) \in \mathring{C}^\infty(\mathbb{R}^n)$. Then the Mellin transform with respect to r is defined by

$$(\mathcal{M}u)(z,\theta) = (2\pi)^{-1/2} \int_0^\infty r^{-iz-1} u(r\theta)dr, \quad z \in \mathbb{C}, \quad \theta \in S^{n-1},$$

and by (4.1.13) and (4.1.14),

$$\int_{-\infty}^{\infty}\int_{S^{n-1}} |\mathcal{M}u(z,\theta)|^2 dt d\theta = \int_0^{\infty}\int_{S^{n-1}} r^{2\tau-1}|u(r\theta)|^2 dr d\theta , \qquad (4.1.15)$$

$$\mathcal{M}(r\,\partial u/\partial r) = iz\mathcal{M}u , \quad z = t + i\tau . \qquad (4.1.16)$$

Let

$$\tilde{L}_2^{\varrho} = L_2(\{\operatorname{Im} z = -\varrho + n/2\} ; L_2(S^{n-1})), \quad \varrho \in \mathbb{R}$$

be the space of square integrable functions on $\operatorname{Im} z = -\varrho + n/2$ with values in $L_2(S^{n-1})$ endowed with the norm

$$\|u\|_{\tilde{L}_2^{\varrho}} = \Big(\int_{-\infty}^{\infty} \int_{S^{n-1}} |u(z,\theta)|^2 dt d\theta \Big)^{1/2} , \quad z = t + i(n/2-\varrho) .$$

For $\varrho \geq 0$ and $\varrho > k$, define

$$\tilde{L}_2^{\varrho,k} = A_2(\Pi_{n/2-\varrho,n/2-k}; L_2(S^{n-1}))$$

to be the space of all functions $u(z,\cdot) \in A_2(\Pi_{n/2-\varrho,n/2-k})$ having values in $L_2(S^{n-1})$ with norm

$$\|u\|_{\tilde{L}_2^{\varrho,k}} = \sup_{k < n/2-\tau < \varrho} \Big(\int_{-\infty}^{\infty} \|u(z,\theta)\|^2_{L_2(S^{n-1})} dt \Big)^{1/2}, \quad z = t + i\tau .$$

We set $\tilde{L}_2^{\varrho;\varrho} = \tilde{L}_2^{\varrho}$. Finally, for $l \in \mathbb{N}$, define the space

$$\tilde{L}_{2,l}^{\varrho,k} = \{ u \in \tilde{L}_2^{\varrho,k} : (1+\delta)^{h/2} z^j u \in \tilde{L}_2^{\varrho,k}, \ h+j \leq l \}$$

endowed with the canonical norm. By Proposition 4.1.5(i) and (4.1.7), $\mathcal{M}$ extends to an isomorphism of $L_2^{\varrho}(\mathbb{R}^n)$ onto $\tilde{L}_2^{\varrho}$. Furthermore, $\mathcal{M}$ is an isomorphism of $L_{2,l}^{\varrho}(\mathbb{R}^n)$ onto $\tilde{L}_{2,l}^{\varrho}$. Indeed, (4.1.16) implies that, for any $u \in \mathcal{S}^{\infty}(\mathbb{R}^n)$,

$$\mathcal{M}((1+\delta)^{h/2}(r\,\partial/\partial r)^j u) = (1+\delta)^{h/2}(iz)^j \mathcal{M}u , \qquad (4.1.17)$$

and by Remark 4.1.4 we obtain that (4.1.17) is valid for any $u \in L_{2,l}^{\varrho}$ and $h+j \leq l$. Analogously, using Proposition 4.1.5(ii), we observe that $\mathcal{M}$ is an isomorphism of $L_2^{\varrho,k}(\mathbb{R}^n)$ and $L_{2,l}^{\varrho,k}(\mathbb{R}^n)$ onto $\tilde{L}_2^{\varrho,k}$ and $\tilde{L}_{2,l}^{\varrho,k}$, respectively.

4.2. The operator $r^2\Delta + ar\partial/\partial r + b$

4.2.1. We consider the differential operator A defined in (4.0.1) which is elliptic outside the origin. By (4.1.5), A can be written in spherical coordinates

$$A = (r\partial/\partial r)^2 + (n-2+a)r\partial/\partial r + b - \delta . \tag{4.2.1}$$

We may choose an orthonormal basis of spherical harmonics

$$\{ P_{m,\alpha}(\theta), m \in \mathbb{N}_0, 1 \le \alpha \le \alpha(m) \}$$

in $L_2(S^{n-1})$ (cf. Stein and Weiss [1, Chap. 4]), where

$$\alpha(m) = (2m+n-2)(n+m-3)!/(n-2)!m!, \quad \delta P_{m,\alpha} = m(m+n-2)P_{m,\alpha} .$$

For any $u \in L_2(\Omega_R)$, $0 < R \le \infty$, we can write

$$u(x) = \sum_{m \ge 0} \sum_{1 \le \alpha \le \alpha(m)} u_{m,\alpha}(r)P_{m,\alpha}(\theta) , \tag{4.2.2}$$

$$u_{m,\alpha}(r) = \int_{S^{n-1}} u(r\theta)\overline{P_{m,\alpha}(\theta)}d\theta , \tag{4.2.3}$$

where the series (4.2.2) converges in L_2 and its terms are mutually orthogonal. Indeed, Parseval's equality implies

$$\int_{S^{n-1}} |u(r\theta)|^2 d\theta = \sum_{m,\alpha} \int_{S^{n-1}} |u_{m,\alpha}(r)P_{m,\alpha}(\theta)|^2 d\theta =$$

$$= \sum_{m,\alpha} |u_{m,\alpha}(r)|^2$$

almost everywhere in (0,R), and by (4.1.7) and a theorem of B. Levi,

$$\|u\|^2_{0,\Omega_R} = \int_0^R \sum_{m,\alpha} |u_{m,\alpha}(r)|^2 r^{n-1}dr = \sum_{m,\alpha} \int_0^R |u_{m,\alpha}(r)|^2 r^{n-1}dr$$

$$= \sum_{m,\alpha} \| u_{m,\alpha} P_{m,\alpha} \|^2_{0,\Omega_R} . \tag{4.2.4}$$

In the sequel we use the equivalent norm

$$|u|_{0,2,\Omega_R} = \sum_{i+j \le 2} \| (1+\delta)^{j/2}(r\partial/\partial r)^i u \|_{0,\Omega_R} \tag{4.2.5}$$

in $L_{2,2}(\Omega_R)$; see 4.1.2. With A we associate the following sequence of Euler operators on (0,R):

$$A_m = (rd/dr)^2 + (n-2+a)rd/dr + b - m(m+n-2), m \in \mathbb{N}_0 . \tag{4.2.6}$$

The following lemma enables us to use the results on Euler equations in the study of the solvability properties of the equation

$$Au = f, \; f = \sum_{m,\alpha} f_{m,\alpha}(r)P_{m,\alpha}(\theta) \in L_2(\Omega_R) \; . \tag{4.2.7}$$

Lemma 4.2.1. (i) If $u \in L_{2,2}(\Omega_R)$ is a solution of (4.2.7), then each function $u_{m,\alpha}$ defined in (4.2.3) is a solution of the equation

$$A_m u_{m,\alpha} = f_{m,\alpha} \tag{4.2.8}$$

and belongs to $L_{2,2}^{(1-n)/2}(0,R)$.

(ii) Conversely, if $u_{m,\alpha} \in L_{2,2}^{(1-n)/2}(0,R)$ $(m \in \mathbb{N}_0, 1 \le \alpha \le \alpha(m))$ are solutions of (4.2.8) satisfying

$$\sum_{m,\alpha} \{ \| u_{m,\alpha} \|^2_{L_{2,2}^{(1-n)/2}(0,R)} + m^4 \| u_{m,\alpha} \|^2_{L_2^{(1-n)/2}(0,R)} + m^2 \| r du_{m,\alpha}/dr \|^2_{L_2^{(1-n)/2}(0,R)} \} < \infty \quad , \tag{4.2.9}$$

then (4.2.2) is a solution of (4.2.7) which belongs to $L_{2,2}(\Omega_R)$.

Proof. (i) Let $u \in L_{2,2}(\Omega_R)$. We show, for example, that

$$r\partial u/\partial r = \sum_{m,\alpha} (r du_{m,\alpha}/dr)P_{m,\alpha} \, .$$

Let $\sum u'_{m,\alpha} P_{m,\alpha}$ be the spherical harmonics expansion of $r\partial u/\partial r$. For any fixed m and α, we choose $v = wP_{m,\alpha}$, where $w \in C_0^\infty(0,R)$ is arbitrary. Since ${}^t(r\partial/\partial r) = -r\partial/\partial r - n$ is the transpose of $r\partial/\partial r$, we obtain

$$\langle r\partial u/\partial r, v\rangle = \int_0^R u'_{m,\alpha}(r)w(r)r^{n-1}dr = \langle u, {}^t(r\partial/\partial r)v\rangle$$

$$= \langle u_{m,\alpha} P_{m,\alpha}, {}^t(r\partial/\partial r)wP_{m,\alpha}\rangle = \int_0^R u_{m,\alpha}\, {}^t(rd/dr)wr^{n-1}dr,$$

where ${}^t(rd/dr) = -rd/dr-1$. Since $r^{1-n}C_0^\infty(0,R) = C_0^\infty(0,R)$, we have $u'_{m,\alpha} = r du_{m,\alpha}/dr$ Q.E.D.

Analogously one can verify that

$$(r\partial u/\partial r)^2 u = \sum_{m,\alpha} (rd/dr)^2 u_{m,\alpha} P_{m,\alpha} \, , \quad \delta u = \sum_{m,\alpha} m(m+n-2)u_{m,\alpha} P_{m,\alpha} \, .$$

To summarize, we see that $u_{m,\alpha} \in L_{2,2}^{(1-n)/2}(0,R)$ in view of (4.2.4) and $Au = \sum (A_m u_{m,\alpha})P_{m,\alpha}$ by virtue of (4.2.1), which gives (4.2.8).

(ii) Let $u_{m,\alpha} \in L_{2,2}^{(1-n)/2}(0,R)$ be solutions of (4.2.8) such that (4.2.9) holds. By (4.2.9) and (4.2.4),

$$u = \sum_{m,\alpha} u_{m,\alpha} P_{m,\alpha} , \; u' = \sum_{m,\alpha} (rdu_{m,\alpha}/dr)P_{m,\alpha} \in L_2(\Omega_R) .$$

We prove that

$$\langle u',v\rangle = \langle u, {}^t(r\partial/\partial r)v\rangle , \; v \in C_0^\infty(\Omega_R) . \tag{4.2.10}$$

By (4.2.3), we have

$${}^t(r\partial/\partial r)v = \sum_{m,\alpha} (-rd/dr-n)v_{m,\alpha} P_{m,\alpha} , \; v_{m,\alpha} = \int_{S^{n-1}} v\overline{P_{m,\alpha}}\, d\theta ,$$

and since the set of all functions in $C^\infty[0,R]$ which are flat at the origin is dense in $L_{2,2}^{(1-n)/2}(0,R)$ (cf. the proof of Lemma 4.1.2),

$$\langle u',v\rangle = \sum_{m,\alpha} \int_0^R (rdu_{m,\alpha}/dr)v_{m,\alpha} r^{n-1}dr$$

$$= \sum_{m,\alpha} \int_0^R u_{m,\alpha} {}^t(rd/dr)v_{m,\alpha} r^{n-1}dr = \langle u, {}^t(r\partial/\partial r)v\rangle ,$$

hence (4.2.10). Thus $u' = r\partial u/\partial r$. Analogously,

$$(1+\delta)^{1/2}u, (r\partial/\partial r)^2u, \delta u, (1+\delta)^{1/2} r\partial u/\partial r \in L_2(\Omega_R)$$

and

$$Au = \sum_{m,\alpha} (A_m u_{m,\alpha})P_{m,\alpha} = \sum_{m,\alpha} f_{m,\alpha} P_{m,\alpha} = f. \; \square$$

The characteristic equation of A_m is

$$\mu^2+(n-2+a)\mu + b-m(m+n-2) = 0$$

which has the roots

$$\mu_j(m) = (2-n-a)/2-(-1)^j\{(2-n-a)^2/4-b+m(m+n-2)\}^{1/2}, j=1,2. \tag{4.2.11}$$

Moreover,

$$\mu_1(m)/m \to 1, \; \mu_2(m)/m \to -1 \; (m\to\infty) . \tag{4.2.12}$$

It follows from Lemma 1.2.2 that the homogeneous equation $A_m v = 0$ has the basis of solutions

$$v_m^1(r)=r^{\mu_1(m)}, \; v_m^2(r) = \begin{cases} r^{\mu_2(m)} & \text{if } \mu_1(m)\neq\mu_2(m), \\ r^{\mu_1(m)}\ln r & \text{if } \mu_1(m)=\mu_2(m) \end{cases} \tag{4.2.13}$$

on (0,R). Set

$$\mathbb{N}_j = \{ m \in \mathbb{N}_0 : \operatorname{Re} \mu_j(m) > -n/2 \}, \; j=1,2.$$

Note that $\mathbb{N}_2$ is finite in view of (4.2.12). We are now ready to prove a theorem on the solvability properties of equation (4.2.7).

<u>Theorem 4.2.2.</u> (i) Under the hypothesis

$$\operatorname{Re} \mu_j(m) \neq -n/2, \; m \in \mathbb{N}_0, \; j=1,2, \tag{4.2.14}$$

the operator $A \in L(L_{2,2}(\mathbb{R}^n), L_2(\mathbb{R}_n))$ is invertible.

(ii) If $R < \infty$ and (4.2.14) is satisfied, then $A \in L(L_{2,2}(\Omega_R), L_2(\Omega_R))$ is surjective with infinite kernel index. More precisely, the function (4.2.2) belongs to the kernel of A in $L_{2,2}(\Omega_R)$ if and only if

$$\begin{aligned} u_{m,\alpha} &= c^1_{m,\alpha} v^1_m + c^2_{m,\alpha} v^2_m, \; c^j_{m,\alpha} \in \mathbb{C} \; (j=1,2), \\ c^j_{m,\alpha} &= 0 \text{ when } m \in \mathbb{N}_0 \setminus \mathbb{N}_j ; \end{aligned} \tag{4.2.15}$$

$$\sum_{m,\alpha} |c^1_{m,\alpha}|^2 m^3 R^{2m} < \infty . \tag{4.2.16}$$

(iii) If (4.2.14) is violated, then $A \in L(L_{2,2}(\Omega_R), L_2(\Omega_R))$, $0 < R \leq \infty$, is not normally solvable.

<u>Proof.</u> 1. Assume (4.2.14). We show that (4.2.7) has a solution in $L_{2,2}(\Omega_R)$ for every right-hand side. The Euler operator (4.2.6) can be written

$$A_m = A_{m2} A_{m1} \, , \; A_{mj} = r d/dr - \mu_j(m) . \tag{4.2.17}$$

By Lemma 1.2.5, Corollary 1.2.6 and Remark 1.2.7, the equations (4.2.8) have solutions $u_{m,\alpha} \in L^{(1-n)/2}_{2,2}(0,R)$ which satisfy

$$\| u_{m,\alpha} \|_{L^{(1-n)/2}_{2,2}(0,R)} \leq c_1(m) c_2(m) \| f_{m,\alpha} \|_{L^{(1-n)/2}_2(0,R)} ,$$

$$\| r du_{m,\alpha}/dr \|_{L^{(1-n)/2}_2(0,R)} \leq c_1(m) d_2(m) \| f_{m,\alpha} \|_{L^{(1-n)/2}_2(0,R)} ,$$

$$\| u_{m,\alpha} \|_{L^{(1-n)/2}_2(0,R)} \leq d_1(m) d_2(m) \| f_{m,\alpha} \|_{L^{(1-n)/2}_2(0,R)} ,$$

where $d_j(m) = |\operatorname{Re} \mu_j(m) + n/2|^{-1}$ and $c_j(m) = d_j(m)(1+|\mu_j(m)|)+1$. Setting $u = \sum u_{m,\alpha} P_{m,\alpha}$, these estimates together with (4.2.4), (4.2.12) and (4.2.14) imply that the left-hand side of (4.2.9) can be estimated by $c \|f\|^2_{0,\Omega_R}$. In view of Lemma 4.2.1 (ii), u is the desired solution of (4.2.7).

2. We now describe the kernel of A. Suppose (4.2.2) belongs to $L_{2,2}(\Omega_R)$ and satisfies the equation $Au = 0$. By Lemma 4.2.1 (i), $u_{m,\alpha}$ belongs to $L^{(1-n)/2}_{2,2}(0,R)$ and satisfies $A_m u_{m,\alpha} = 0$ for any m and α. If $R = \infty$, then it follows from Remark 1.2.7 that $u_{m,\alpha} = 0$ for all m and α, hence $u = 0$. Let $R < \infty$. By Lemma 1.2.2, $u_{m,\alpha}$ takes the form (4.2.15). Moreover, by $\delta u \in L_2(\Omega_R)$, (4.2.12) and (4.2.14),

$$\|\delta u\|^2_{0,\Omega_R} \geq \sum_{m,\alpha} |c^1_{m,\alpha}|^2 m^2(m+n-2)^2 \int_0^R r^{2\mathrm{Re}\,\mu_1(m)+n-1} dr$$

$$\geq c \sum_{m,\alpha} m^4 |c^1_{m,\alpha}|^2 m^{-1} R^{2m}$$

which implies (4.2.16). Conversely, if (4.2.16) holds, then with $u = \sum u_{m,\alpha} P_{m,\alpha}$ and $u_{m,\alpha}$ defined by (4.2.15) we get

$$\|\delta u\|^2_{0,\Omega_R} \leq 2 \Big\{ \sum_{j=1,2} \sum_{m \in \mathbb{N}_j,\alpha} |c^j_{m,\alpha}|^2 m^2 (m+n-2)^2 \int_0^R |v^j_m|^2 r^{n-1} dr \Big\}$$

$$\leq c \sum_{m,\alpha} m^3 R^{2m} |c^1_{m,\alpha}|^2 + c_1 < \infty ,$$

since the set $\mathbb{N}_2$ is finite. Similarly, the other terms in (4.2.5) are finite so that $u \in L_{2,2}(\Omega_R)$ and $Au = 0$. The proof of (i) and (ii) is complete.

3. It remains to show (iii). Let $\mathrm{Re}\,\mu_j(m) = -n/2$ for some $j=1,2, m \in \mathbb{N}_0$, and choose some α, $1 \leq \alpha \leq \alpha(m)$. Then $A_m \in L(L^{(1-n)/2}_{2,2}(0,R), L^{(1-n)/2}_2(0,R))$ is not normally solvable. This follows from Theorem 1.3.1 (ii) when $R < \infty$ and from Example 3.4.4 (for p=2, k=0 and the Euler operator $r^{(n-1)/2} A_m r^{(1-n)/2}$) when $R = \infty$.

Therefore one may choose a sequence $\{u^N\} \subset L^{(1-n)/2}_{2,2}(0,R)$ and an element $f \in L^{(1-n)/2}_2$ such that $A_m u^N \to f$ $(N \to \infty)$ in $L^{(1-n)/2}_2(0,R)$ and $f \bar{\in} A_m(L^{(1-n)/2}_{2,2}(0,R))$. Setting $g = f P_{m,\alpha}$, we obtain $g \in L_2(\Omega_R)$, $v^N = u^N P_{m,\alpha} \in L_{2,2}(\Omega_R)$ and $Av^N \to g$ $(N \to \infty)$ in $L_2(\Omega_R)$, but $g \bar{\in} A(L_{2,2}(\Omega_R))$. Hence A cannot be normally solvable. □

4.2.2. We now study the operator (4.o.1) in $C^\infty(\bar{\Omega}_R)$, $0 < R < \infty$, and in the subspace of flat functions

$$C^{\infty,0}(\bar{\Omega}_R) = \{u \in C^\infty(\bar{\Omega}_R) : D^\alpha u(0) = 0,\ \alpha \in \mathbb{N}_0^n\}.$$

__Theorem 4.2.3.__ For any $a,b \in \mathbb{C}$, A is not normally solvable in $C^{\infty,0}(\bar{\Omega}_R)$.

__Proof.__ Choose $\chi \in C_0^\infty(0,\infty)$ such that $\chi(r) = 0$ for $r \leq 1/2$ and $r \geq 4$ and $\chi(r) = 1$ for $1 \leq r \leq 3$. For any $m \in \mathbb{N}_0$ and $\alpha, 1 \leq \alpha \leq \alpha(m)$, we set

$$f_{m,\alpha}(r) = (rd/dr - \mu_2(m))\{r^{i\,\mathrm{Im}\,\mu_1(m)} g_{m,\alpha}(r)\}, \tag{4.2.18}$$

where $\mu_j(m)$ are defined by (4.2.11) and $g_{m,\alpha} = 2^{-m}\chi(mr)(\alpha(m))^{-1/2}$. Let $f = \sum_{m,\alpha} f_{m,\alpha} P_{m,\alpha}$. We first verify that $f \in C^{\infty,0}(\bar{\Omega}_R)$.

Obviously, $f_{m,\alpha} P_{m,\alpha} \in C^{\infty,0}(\bar{\Omega}_R)$ for all m and α. It remains to show that the series $\sum \Delta^k(f_{m,\alpha} P_{m,\alpha})$ converges in $L_2(\mathbb{R}^n)$ for all $k \in \mathbb{N}_0$, since $(1-\Delta)^k$ is an isomorphism of $W_2^{2k}(\mathbb{R}^n)$ onto $L_2(\mathbb{R}^n)$ and $C^\infty(\bar{\Omega}_R)$ is the projective limit of the spaces $W_2^k(\Omega_R)$. By (4.2.12) and (4.2.18), we obtain $\Delta^k f_{m,\alpha} P_{m,\alpha} = h_{m,\alpha} P_{m,\alpha}$ with some function $h_{m,\alpha}$ which can be estimated by

$$c2^{-m}(\alpha(m))^{-1/2} m^{2k+1} r^{-2k} \sum_{0 \leq j \leq 2k+1} d^j \chi(mr)/dr^j$$

$$\leq c_1 2^{-m}(\alpha(m))^{-1/2} m^{6k+2} \|\chi\|_{C^{4k+1}[0,4]}, \quad 0 \leq r \leq 4,$$

where c and c_1 are independent of m, α and r. Consequently,

$$\sum_{m,\alpha} \|h_{m,\alpha} P_{m,\alpha}\|^2_{0,\mathbb{R}^n} \leq c_2 \sum_{m \geq 0} m^{12k+4} 2^{-2m} < \infty \qquad \text{Q.E.D.}$$

Let $f_N = \sum_{m \leq N,\alpha} f_{m,\alpha} P_{m,\alpha}$. We have just proved that f_N converges to f in $C^{\infty,0}(\bar{\Omega}_R)$ as $N \to \infty$. By (4.2.17) and (4.2.18), the equation $A_m u_{m,\alpha} = f_{m,\alpha}$ has the solution

$$u_{m,\alpha}(r) = r^{\mu_1(m)} \int_0^r \varrho^{-\mathrm{Re}\,\mu_1(m)-1} g_{m,\alpha}(\varrho)\,d\varrho \tag{4.2.19}$$

(cf. the proof of Lemma 1.2.5) which clearly belongs to $C^{\infty,0}[0,R]$. Since

$$u_N = \sum_{m \leq N,\alpha} u_{m,\alpha} P_{m,\alpha} \in C^{\infty,0}(\bar{\Omega}_R)$$

is a solution of $Au = f_N$, f belongs to the closure of the range of A in $C^{\infty,0}(\bar{\Omega}_R)$.

Suppose now $Au = f$ has a solution $u \in C^{\infty,0}(\bar{\Omega}_R)$, and let $u_{m,\alpha}$ be the functions (4.2.3). Then $u_{m,\alpha} \in C^{\infty,0}[0,R]$ and $A_m u_{m,\alpha} = f_{m,\alpha}$ for any m and α. But Theorem 2.4.6 implies that (4.2.19) is the unique solution of this equation in $C^{\infty,0}[0,R]$. Finally, we show that

$$\|u\|^2_{0,\Omega_R} = \sum_{m,\alpha} \int_0^R |u_{m,\alpha}|^2 r^{n-1} dr = \infty, \qquad (4.2.20)$$

hence a contradiction which completes the proof of the theorem. Using (4.2.18), (4.2.19) and (4.2.12), we get the estimates

$$\begin{aligned} |u_{m,\alpha}(r)| &= r^{\mathrm{Re}\,\mu_1(m)} \int_0^r \varrho^{-\mathrm{Re}\,\mu_1(m)-1} g_{m,\alpha}(\varrho)\,d\varrho \\ &\ge 2^{-m}(\alpha(m))^{-1/2}(\mathrm{Re}\,\mu_1(m))^{-1}(3/m)^{\mathrm{Re}\,\mu_1(m)}\{m^{\mathrm{Re}\,\mu_1(m)} - (m/3)^{\mathrm{Re}\,\mu_1(m)}\} \\ &\ge c(\alpha(m))^{-1/2} m^{-1} \gamma^m \end{aligned} \qquad (4.2.21)$$

with some $\gamma > 1$ for all $r \ge 3/m$, α and sufficiently large m. (4.2.21) implies

$$\sum_{m \ge \bar{m},\alpha} \int_{3/m}^{4/m} |u_{m,\alpha}|^2 r^{n-1} dr \ge c_1 \sum_{m \ge \bar{m}} m^{-3}\gamma^m = \infty$$

for some $\bar{m} \in \mathbb{N}$, hence (4.2.20). □

Finally, we use Theorem 4.2.3 to derive a criterion for the normal solvability of (4.0.1) in $C^\infty(\bar{\Omega}_R)$.

<u>Theorem 4.2.4.</u> A is normally solvable in $C^\infty(\bar{\Omega}_R)$ if and only if there exists $j \in \mathbb{N}_0$ such that $a = -4j$ and $b = 2j(2j-n+2)$. (4.2.22)

<u>Proof.</u> We first show that, for $v = \sum v_{m,\alpha} P_{m,\alpha} \in C^\infty(\bar{\Omega}_R)$, the functions $\tilde{v}_{m,\alpha}$ defined by

$$\tilde{v}_{m,\alpha}(r^2) = r^{-m} v_{m,\alpha}(r), \quad 0 \le r \le R$$

belong to $C^\infty[0,R^2]$. Every homogeneous polynomial P of degree k can be represented in the form $P = P_0 + r^2 P_1 + \ldots + r^{2l} P_l$, where P_j are homogeneous harmonic polynomials of degree $k-2j$ $(j=0,\ldots,l)$; see Stein and Weiss [1, Chap. IV.2]. The restriction of P_j to S^{n-1} is a linear combination of the spherical harmonics $P_{k-2j,\alpha}$, $1 \le \alpha \le \alpha(k-2j)$. Therefore, using the Taylor expansion of v at the origin and the orthogonality properties of $P_{m,\alpha}$, we obtain the assertion from (4.2.3).

Next, we prove the following assertion: If

$$\mu_1(m)/2 - m \in \mathbb{N}_0 \quad \text{for an infinite set of indices } m, \tag{4.2.23}$$

then (4.2.22) holds.

Indeed, (4.2.23) implies that $(2-n-a)^2-4b+4m^2+4m(n-2)$ is the square of an integer for arbitrarily large m, hence

$$(2-n-a)^2-4b = (n-2)^2, \quad \mu_1(m) = m-a/2 .$$

Furthermore, (4.2.23) implies $a = -4j$ with some $j \in \mathbb{N}_0$, and then

$$b = a(a+2n-4)/4 = 2j(2j-n+2) \qquad \text{Q.E.D.}$$

Assume now that (4.2.22) is violated, and let $f = \sum f_{m,\alpha} P_{m,\alpha}$ be the function constructed in the proof of Theorem 4.2.3. Then f belongs to the closure of the range of A in $C^\infty(\bar{\Omega}_R)$. Suppose the equation $Av = f$ has a solution $v = \sum v_{m,\alpha} P_{m,\alpha}$ in $C^\infty(\bar{\Omega}_R)$. Then $A_m v_{m,\alpha} = f_{m,\alpha}$, so $v_{m,\alpha}$ has the representation

$$v_{m,\alpha} = u_{m,\alpha} + c^1_{m,\alpha} v^1_m + c^2_{m,\alpha} v^2_m , \quad c^j_{m,\alpha} \in \mathbb{C}$$

on (0,R), where v^j_m are defined by (4.2.13) and $u_{m,\alpha} \in C^{\infty,0}[0,R]$ is given by (4.2.19). Since the corresponding functions $\tilde{v}_{m,\alpha}$ defined above belong to $C^\infty[0,R^2]$, $\mu_2(m) \to -\infty$ $(m \to \infty)$ and $\mu_1(m)/2-m \notin \mathbb{N}_0$ for all sufficiently large m, we obtain $c^1_{m,\alpha} = c^2_{m,\alpha} = 0$ for those m. Then (4.2.20) yields $v \notin L_2(\Omega_R)$, hence a contradiction.

It remains to prove that A is normally solvable if (4.2.22) holds. Then $A = r^{2j+2} \Delta r^{-2j}$, $j \in \mathbb{N}_0$. Since Δ is a surjective map in $C^\infty(\bar{\Omega}_R)$ (cf. e.g. Lions and Magenes [1, Chap. 2]) and the multiplication operator r^{2j+2} is normally solvable in $C^\infty(\bar{\Omega}_R)$ (see e.g. Hörmander [1]), A is normally solvable there. □

4.3. A class of elliptic operators degenerating at the origin

4.3.1. We consider the differential operator

$$A = A(x,D) = \sum_{|\alpha|=|\beta|\le 1} c_{\alpha\beta}\, x^\beta D^\alpha , \quad c_{\alpha\beta} \in \mathbb{C} \tag{4.3.1}$$

in $\mathbb{R}^n$ and assume throughout this section that A is elliptic in $\mathbb{R}^n \setminus \{0\}$:

$$p_l(x,\xi) = \sum_{|\alpha|=|\beta|=l} c_{\alpha\beta}\, x^{\beta} i^{|\alpha|} \xi^{\alpha} \neq 0, \quad x,\xi \in \mathbb{R}^n \setminus \{0\} \ . \qquad (4.3.2)$$

A can be written in spherical coordinates (see (4.1.3), (4.1.4))

$$A(\theta,D_\theta,r\partial/\partial r) = \sum_{0\le j\le l} A_j(\theta,D_\theta)(r\partial/\partial r)^{l-j}, \qquad (4.3.3)$$

where $A_j(\theta,D_\theta)$ are differential operators of order j on S^{n-1} with C^∞ coefficients. With (4.3.3) we associate the operator function

$$\mathcal{A}_z = \sum_{0\le j\le l} A_j(\theta,D_\theta)(iz)^{l-j}, \quad z\in\mathbb{C} \quad . \qquad (4.3.4)$$

$\mathcal{A}_z$ is obviously an analytic operator function in $\mathbb{C}$ with values in $L(W_2^l(S^{n-1}),L_2(S^{n-1}))$. Passing to spherical coordinates, we obtain a C^∞ diffeomorphism

$$\psi : \mathbb{R}^n \setminus \{0\} \to S^{n-1}\times (0,\infty)$$

which generates a C^∞ diffeomorphism (see Narasimhan [1])

$$\psi^* : T^*(S^{n-1})\times \mathbb{R} \to T^*(\mathbb{R}^n)|_{S^{n-1}} = S^{n-1}\times \mathbb{R}^n ,$$

where $T^*(M)$ denotes the cotangent bundle of a C^∞ manifold M. For $((\theta,\eta),z)\in T^*(S^{n-1})\times\mathbb{R}$, let $a(\theta,\eta,z)$ be the composition of the function p_l defined in (4.3.2) with the diffeomorphism ψ^*. It is a homogeneous polynomial of degree l with respect to the variables (η,z) which takes the form

$$a(\theta,\eta,z) = \sum_{0\le j\le l} a_j(\theta,\eta)(iz)^{l-j} ,$$

where $a_j(\theta,\eta)$, $(\theta,\eta)\in T^*(S^{n-1})$, is the symbol of the differential operator $A_j(\theta,D_\theta)$ on S^{n-1} (for definition, see Narasimhan [1]).

By (4.3.2) there exists a conic neighborhood

$$\Gamma = \{z\in\mathbb{C} : -\varepsilon<\arg z<\varepsilon\,,\ \pi-\varepsilon<\arg z<\pi+\varepsilon\}\,, \quad \varepsilon\in(0,\pi/2)$$

of the real axis such that $\mathcal{A}_z$ is elliptic with parameter z in $S^{n-1}\times\Gamma$:

$$a(\theta,\eta,z) \neq 0, \ (\theta,\eta)\in T^*(S^{n-1}), \ z\in\Gamma\,, \ |z|+|\eta|\neq 0 \ . \qquad (4.3.5)$$

In particular, $\mathcal{A}_z$ is elliptic in the usual sense for any $z\in\mathbb{C}$, since (4.3.5) for $z=0$ implies $a_l(\theta,\eta)\neq 0$, $(\theta,\eta)\in T^*(S^{n-1})$, $\eta\neq 0$.

A complex number z_0 is called eigenvalue of the operator function (4.3.4) if there exists $u_0\in W_2^l(S^{n-1})$, $u_0\neq 0$, such that $\mathcal{A}_{z_0}u_0 = 0$. Then

$u_0 \in C^{\infty}(S^{n-1})$ in view of the elliptic regularity. The following proposition which is basic for our further considerations follows immediately from (4.3.5) and the results in Agranovič and Višik [1].

Proposition 4.3.1. For any $z \in \mathbb{C}$, $\mathcal{A}_z \in L(W_2^l(S^{n-1}), L_2(S^{n-1}))$ is a Fredholm operator with index 0, and there exists only a discrete set $\{z_m\}$ of eigenvalues of $\mathcal{A}_z$. Furthermore, there exist numbers $c, M > 0$ such that the inverse $\mathcal{A}_z^{-1}$ of $\mathcal{A}_z$ exists and satisfies the estimate

$$|z|^l \|\mathcal{A}_z^{-1} f\|_{0,S^{n-1}} + \|\mathcal{A}_z^{-1} f\|_{l,S^{n-1}} \le c \|f\|_{0,S^{n-1}} \qquad (4.3.6)$$

for all $f \in L_2(S^{n-1})$ and $z \in \Gamma$, $|z| \ge M$.

4.3.2. We first consider the operator (4.3.1) as a continuous linear map of $L^{\rho}_{2,l}(\mathbb{R}^n)$ into $L^{\rho}_2(\mathbb{R}^n)$, $\rho \ge 0$ (see 4.1.3).

Theorem 4.3.2. Under the hypothesis

$$\operatorname{Im} z_m \neq -\rho + n/2 \quad \text{for all eigenvalues of } \mathcal{A}_z , \qquad (4.3.7)$$

the operator $A \in L(L^{\rho}_{2,l}(\mathbb{R}^n), L^{\rho}_2(\mathbb{R}^n))$ is invertible; otherwise it is not normally solvable.

Proof. 1. Consider $\mathcal{A}_z$ as a continuous linear map of $\tilde{L}^{\rho}_{2,l}$ into $\tilde{L}^{\rho}_2$ (see 4.1.4). Using Remark 4.1.4, we obtain

$$\mathcal{M}\{A_j(\theta, D_\theta)(r\partial/\partial r)^{l-j}u\} = A_j(\theta, D_\theta)(iz)^{l-j}\mathcal{M}u, u \in L^{\rho}_{2,l} .$$

Thus we have the commutative diagram

$$\begin{array}{cccc} A & : & L^{\rho}_{2,l} \longrightarrow & L^{\rho}_2 \\ & & \downarrow \mathcal{M} \quad & \downarrow \mathcal{M} \\ \mathcal{A}_z & : & \tilde{L}^{\rho}_{2,l} \longrightarrow & \tilde{L}^{\rho}_2 \end{array} , \qquad (4.3.8)$$

where the Mellin transform $\mathcal{M}$ is an isomorphism between the corresponding spaces. Assume that (4.3.7) holds. Then, by Proposition 4.3.1, estimate (4.3.6) holds for all z on the line $\operatorname{Im} z = -\rho + n/2$. Let $f \in \tilde{L}^{\rho}_2$. Setting

$$u(z, \theta) = \mathcal{A}_z^{-1} f(z, \theta), \quad \theta \in S^{n-1}, \ \operatorname{Im} z = -\rho + n/2 \qquad (4.3.9)$$

and using the interpolation inequalities

$$\|v\|_{k,S^{n-1}} \le c \|v\|^{1-k/l}_{0,S^{n-1}} \|v\|^{k/l}_{l,S^{n-1}}, \ v \in W_2^l(S^{n-1}), \ k=1,\dots,l-1 \qquad (4.3.10)$$

which are an easy consequence of definition (4.1.6), we obtain

$$\sum_{h+j\le 1} \|(1+\delta)^{h/2} z^j u(z,\theta)\|^2_{0,S^{n-1}} \le c\,\{\, |z|^{2l}\|\mathcal{A}_z^{-1} f\|^2_{0,S^{n-1}}$$

$$+\|\mathcal{A}_z^{-1} f\|^2_{0,S^{n-1}}\} \le c\,\|f(z,\theta)\|^2_{0,S^{n-1}}, \ \operatorname{Im} z = -\varrho + n/2 .$$

Integrating the last inequality on the line $\operatorname{Im} z = -\varrho + n/2$, we see that $u \in \tilde{L}^{\varrho}_{2,1}$. Furthermore, $u \in \tilde{L}^{\varrho}_{2,1}$ and $\mathcal{A}_z u = 0$ imply $u = 0$ since $\mathcal{A}_z \in L(W^1_2(S^{n-1}), L_2(S^{n-1}))$ is invertible for almost all z on the line $\operatorname{Im} z = -\varrho + n/2$. Note that condition (4.3.7) is not necessary here. Consequently, for any $f \in \tilde{L}^{\varrho}_2$, (4.3.9) is the unique solution of the equation $\mathcal{A}_z u = f$ in $\tilde{L}^{\varrho}_{2,1}$. Hence, by virtue of (4.3.8), A is invertible.

2. We now show that A is not normally solvable if $\operatorname{Im} z_1 = -\varrho + n/2$ for some eigenvalue z_1 of $\mathcal{A}_z$. By (4.3.8) it suffices to prove that $\mathcal{A}_z \in L(\tilde{L}^{\varrho}_{2,1}, \tilde{L}^{\varrho}_2)$ is not normally solvable. Since $\mathcal{A}_z$ is injective, this can be done by constructing a sequence $\{u_m\} \subset \tilde{L}^{\varrho}_{2,1}$ such that

$$\|\mathcal{A}_z u_m\|_{\tilde{L}^{\varrho}_2} \to 0 , \quad \|u_m\|_{\tilde{L}^{\varrho}_{2,1}} \ge c > 0 \ (m \to \infty) . \qquad (4.3.11)$$

There exists $v_1 \in C^{\infty}(S^{n-1})$, $v_1 \not\equiv 0$, such that $\mathcal{A}_{z_1} v_1(\theta) = 0$. Define a sequence of functions χ_m on the line $\operatorname{Im} z = -\varrho + n/2$ by $\chi_m = m^{1/2}$ for $|z-z_1| \le 1/2m$ and $\chi_m = 0$ for $|z-z_1| > 1/2m$ and let $u_m = \chi_m v_1$. Then we obtain $u_m \in \tilde{L}^{\varrho}_{2,1}$ and

$$\|u_m\|^2_{\tilde{L}^{\varrho}_2} = \|\chi_m\|^2_{L_2(\{\operatorname{Im} z = n/2-\varrho\})} \|v_1\|^2_{0,S^{n-1}} = \|v_1\|^2_{0,S^{n-1}} > 0 ,$$

$$\|\mathcal{A}_z u_m\|_{\tilde{L}^{\varrho}_2} = \|\chi_m(\mathcal{A}_z - \mathcal{A}_{z_1}) v_1\|_{\tilde{L}^{\varrho}_2}$$

$$\le \max_{|z-z_1|\le 1/2m} \|\mathcal{A}_z - \mathcal{A}_{z_1}\|_{L(W^1_2(S^{n-1}), L_2(S^{n-1}))} \|v_1\|_{W^1_2(S^{n-1})} \to 0$$

as $m \to \infty$, hence (4.3.11) Q.E.D. □

We say that a differential operator P with C^{∞} coefficients is locally solvable at a point x if there exists an open set U containing x such that $P(\mathcal{D}'(U)) \supset C^{\infty}_0(U)$. The operator (4.3.1) is of course locally solvable at any point $x \neq 0$ since it is elliptic there. As a consequence of Theorem 4.3.2, we obtain

Corollary 4.3.3. The operator (4.3.1) is locally solvable at the origin

Proof. In view of Proposition 4.3.1 there exists $\varrho \in [0,n/2)$ such that $\mathcal{A}_z$ has no eigenvalue on the line $\operatorname{Im} z = -\varrho + n/2$. Then it follows from Theorem 4.3.2 that, for any $f \in C_0^\infty(\mathbb{R}^n)$, the equation $Au = f$ has a solution $u \in L^\varrho_{2,1}(\mathbb{R}^n)$. (Note that $C_0^\infty \subset L_2^\varrho$ if $\varrho < n/2$.) $\square$

We now extend Theorem 4.3.2 to the spaces $L_2^{\varrho,k}$, $\varrho \geq 0$, $\varrho > k$.

Theorem 4.3.4. Under the hypotheses

$$\operatorname{Im} z_m \neq -\varrho + n/2, -k+n/2 \quad \text{for all eigenvalues of } \mathcal{A}_z, \tag{4.3.12}$$

$A \in L(L^{\varrho,k}_{2,1}(\mathbb{R}^n), L_2^{\varrho,k}(\mathbb{R}^n))$ is an injective Fredholm operator.

Proof. One obtains the commutative diagram (see 4.1.4)

$$\begin{array}{cccc} A & : L^{\varrho,k}_{2,1} & \longrightarrow & L^{\varrho,k}_2 \\ & \downarrow \mathcal{M} & & \downarrow \mathcal{M} \\ \mathcal{A}_z & : \tilde{L}^{\varrho,k}_{2,1} & \longrightarrow & \tilde{L}^{\varrho,k}_2 \end{array} .$$

We have to prove that $\mathcal{A}_z \in L(\tilde{L}^{\varrho,k}_{2,1}, \tilde{L}^{\varrho,k}_2)$ is an injective Fredholm operator. Since the injectivity was already shown in the proof of Theorem 4.3.2, it remains to verify that $\mathcal{A}_z$ has finite deficiency. Let $f \in \tilde{L}_2^{\varrho,k}$ and let $\{z_m : m = 1,\dots,N\}$ be the set of eigenvalues of $\mathcal{A}_z$ lying in the strip

$$\Pi = \{z \in \mathbb{C} \ : -\varrho + n/2 < \operatorname{Im} z < -k+n/2\} .$$

This set is finite by Proposition 4.3.1. Choose mutually disjoint open neighborhoods $U_m \subset \Pi$ of z_m. Then it follows from (4.3.12), Proposition 4.3.1 and (4.3.10) that the function

$$u(z,\theta) = \mathcal{A}_z^{-1} f(z,\theta) \ , \ z \in \Pi \setminus \{z_1,\dots,z_N\} \ , \ \theta \in S^{n-1} \tag{4.3.13}$$

satisfies the estimate

$$\begin{aligned} &\sum_{h+j \leq 1} \|(1+\delta)^{h/2} z^j u(z,\theta)\|^2_{0,S^{n-1}} \leq c\, \|f(z,\theta)\|^2_{0,S^{n-1}} , \\ &z \in \Pi \setminus \bigcup_{1 \leq m \leq N} U_m . \end{aligned} \tag{4.3.14}$$

Furthermore, in every neighborhood U_m the operator function $\mathcal{A}_z^{-1}$ has a Laurent expansion

$$\mathcal{A}_z^{-1} = \sum_{j \geq n(m)} (z-z_m)^j C_j^{(m)} \ , \ C_j^{(m)} \in L(L_2(S^{n-1}), W_2^1(S^{n-1})), \tag{4.3.15}$$

where $n(m) \in \mathbb{Z}$ and $C_j^{(m)}$ are operators with finite-dimensional range when $j < 0$. This follows from Proposition 4.3.1 and general results on analytic operator functions in Seeley [1] or Blekher [1]. Therefore, by (4.3.14), $\mathcal{A}_z u = f$ has a solution in $\tilde{L}_{2,1}^{\varrho,k}$ if and only if the function $u(z,.)$ defined in (4.3.13) is analytic in each set U_m. Expanding $f(z,.)$ into a Taylor series at the point z_m and using (4.3.15), we observe that $u(z,.)$ is analytic in U_m if and only if

$$\sum_{0\leq r\leq j} (\partial/\partial z)^{j-r} C^{(m)}_{n(m)+r} f(z,\theta)/(j-r)!\,|_{z=z_m} = 0\,, \qquad j = 0,\dots,n(m)-1. \tag{4.3.16}$$

It is easily seen that, for $m = 1,\dots,N$, the left-hand sides of (4.3.16) determine a finite set of continuous linear functionals on $\tilde{L}_2^{\varrho,k}$. This completes the proof of the theorem. □

<u>4.3.3.</u> We now study the operator (4.3.1) in Sobolev spaces on $\mathbb{R}^n$ and on bounded domains.

<u>Theorem 4.3.5.</u> Under the hypotheses

$$\operatorname{Im} z_m \neq n/2-k,\ n/2 \qquad \text{for all eigenvalues of } \mathcal{A}_z\,, \tag{4.3.17}$$

$A\in L(W^k_{2,1}(\mathbb{R}^n), W^k_2(\mathbb{R}^n))$ is an injective Fredholm operator.

<u>Proof.</u> It follows from 4.1.3 that the diagram

$$\begin{array}{rcccl} A: & W^k_{2,1} & \longrightarrow & W^k_2 & \\ & \downarrow \mathcal{F} & & \downarrow \mathcal{F} & \\ \hat{A}: & L^{o,-k}_{2,1} & \longrightarrow & L^{o,-k}_2 & \end{array} \tag{4.3.18}$$

is commutative, where the Fourier transform $\mathcal{F}$ is an isomorphism between the corresponding spaces and

$$\hat{A} = \sum_{|\alpha|=|\beta|\leq 1} c_{\alpha\beta}\ (-1)^{|\alpha|}\, D^{\beta}\, x^{\alpha}\ . \tag{4.3.19}$$

(4.3.19) is an operator of the form (4.3.1) which is also elliptic in $\mathbb{R}^n\setminus\{0\}$ by (4.3.2). In analogy to (4.3.4), with $\hat{A}$ we can associate an operator function $\hat{\mathcal{A}}_z$ for which Proposition 4.3.1 is valid again. Next, we verify that $\mathcal{A}_z$ has the eigenvalue $z_0\in\mathbb{C}$ if and only if $-z_0+in$ is an eigenvalue of $\hat{\mathcal{A}}_z$: Suppose first that $iz_0\bar{\in}\mathbb{Z}$. Let $\mathcal{A}_{z_0}u_0(\theta) = 0$ with $u_0\in C^\infty(S^{n-1})$, $u_0\neq 0$. Then (4.3.3) implies $Ar^{iz_0}u_0 = 0$, and the Fourier transform of the tempered distribution $r^{iz_0}u_0$ takes the form $r^{-iz_0-n}\hat{u}_0$, where $\hat{u}_0\in C^\infty(S^{n-1})$ (cf. Eskin

[1, § 1.9]). Thus we obtain $\hat{A}r^{-iz_o-n}\hat{u}_o = 0$, hence $\hat{\mathcal{A}}_{-z_o+in}\hat{u}_o = 0$ and $\hat{u}_o \neq 0$. Conversely, if $-z_o+in$ is an eigenvalue of $\hat{\mathcal{A}}_z$, then z_o is an eigenvalue of $\mathcal{A}_z$. The case $iz_o \in \mathbb{Z}$ may be reduced to the first.

Finally, since (4.3.17) is equivalent to condition (4.3.12) (with $\varrho = 0$) for the operator $\hat{A}$, Theorem 4.3.4 yields that $\hat{A} \in L(L_{2,1}^{o,-k}, L_2^{o,-k})$ is an injective Fredholm operator. In view of (4.3.18) the theorem is proved. □

<u>Remark 4.3.6.</u> Similarly as in Theorem 4.3.2 one can prove that $A \in L(W_{2,1}^k(\mathbb{R}^n), W_2^k(\mathbb{R}^n))$ is not normally solvable if (4.3.17) is violated.

Henceforth let $\Omega \subset \mathbb{R}^n$ be a bounded domain as in 4.1.1 which contains the origin. We have seen in Sec. 4.2 that the operator (4.3.1) is not surjective in the space of flat C^∞ functions, in general. However, it is possible to prove a certain analogue of Theorem 1.2.4. Similarly as in the one-dimensional case (see Lemma 1.1.1) one can verify that $A \in L(\mathring{W}_{2,1}^k(\Omega), \mathring{W}_2^k(\Omega))$.

<u>Theorem 4.3.7.</u> Under the condition

$$\operatorname{Im} z_m \neq n/2-k \quad \text{for all eigenvalues of } \mathcal{A}_z, \tag{4.3.2o}$$

the operator $A \in L(\mathring{W}_{2,1}^k(\Omega), \mathring{W}_2^k(\Omega))$ is surjective.

As a consequence of this theorem, we have

<u>Corollary 4.3.8.</u> Assume (4.3.2o). Then the equation

$$Au = f, \quad f \in W_2^k(\Omega) \tag{4.3.21}$$

has a solution $u \in W_{2,1}^k(\Omega)$ if and only if the linear system

$$(D^\alpha A\hat{u})(0) = (D^\alpha f)(0), \quad |\alpha| < k-n/2$$

in the unknowns $(D^\alpha \hat{u})(0)$, $|\alpha| < k-n/2$, is solvable, where $\hat{u}$ denotes a formal power series in the indeterminates $x_1, \ldots, x_n$. Thus $A \in L(W_{2,1}^k(\Omega), W_2^k(\Omega))$ is a Φ_--operator if (4.3.2o) holds.

<u>Proof of Theorem 4.3.7.</u> For $k = 0$, the assertion follows from Theorem 4.3.2 by extending the right-hand side of (4.3.21) to a function in $L_2(\mathbb{R}^n)$. Let $k > 0$. First, we prove the relation

$$\dim W_2^k(\Omega)/A(W_{2,1}^k(\Omega)) < \infty . \tag{4.3.22}$$

By Theorem 9.1 in Lions and Magenes [1, Chap. 2], there is a continuous extension operator $p \in L(W_2^k(\Omega), W_2^k(\mathbb{R}^n))$ such that $pu|_\Omega = u$, $u \in W_2^k(\Omega)$. Choose $\chi \in C_0^\infty(\mathbb{R}^n)$ such that $\chi = 1$ in Ω. Consider the equation

$$\hat{A}\hat{u} = \hat{f} \tag{4.3.23}$$

with $\hat{A}$ defined in (4.3.19) and $\hat{f} = \mathcal{F}(\chi pf)$. Since χpf has support in a fixed compact set, one obtains the estimate

$$\| \hat{f} \|_{L_2^{\varrho,k}(\mathbb{R}^n)} \le c(\varrho) \| f \|_{W_2^k(\Omega)} , \; f \in W_2^k(\Omega) \tag{4.3.24}$$

for all $\varrho \in [0, n/2)$. Furthermore, one may choose ϱ such that, in addition, (4.3.12) for $\hat{A}$ is satisfied. By Theorem 4.3.4, equation (4.3.23) has a solution $\hat{u} \in L_{2,1}^{\varrho,k}(\mathbb{R}^n)$ if $\hat{f}$ satisfies a finite number of solvability conditions in $L_2^{\varrho,k}(\mathbb{R}^n)$. Since $L_{2,1}^{\varrho,k} \subset L_{2,1}^{0,k}$, (4.3.18) and (4.3.24) imply that equation (4.3.21) has the solution $\mathcal{F}^{-1}\hat{u}|_\Omega \in$ $\in W_{2,1}^k(\Omega)$ if f satisfies finitely many solvability conditions in $W_2^k(\Omega)$, where $\mathcal{F}^{-1}$ denotes the inverse Fourier transform. Hence (4.3.22) is proved.

Further, in view of (4.3.22) and the relations $\dim W_2^k/\mathring{W}_2^k < \infty$, $\dim W_{2,1}^k/\mathring{W}_{2,1}^k < \infty$ we obtain

$$A(\mathring{W}_{2,1}^k(\Omega)) \dotplus N = \mathring{W}_2^k(\Omega), \; \dim N < \infty . \tag{4.3.25}$$

Since $\mathring{C}^\infty(\Omega)$ is dense in $\mathring{W}_2^k(\Omega)$ by Lemma 4.1.1, we can assume that the space N in (4.3.25) is the span of certain functions $f_j \in \mathring{C}^\infty(\Omega)$, $j=1,\dots,q = \dim N$; cf. Przeworska-Rolewicz and Rolewicz [1, Chap. B.III, Lemma 2.3]. Choose $\varepsilon > 0$ such that

$$\varepsilon\Omega \cap \operatorname{supp} f_j = \emptyset \quad (j=1,\dots,q), \quad \varepsilon\Omega = \{ \varepsilon x : x \in \Omega \} .$$

Now we prove that the equation

$$Au = f, \; f \in \mathring{W}_2^k(\varepsilon\Omega) \tag{4.3.26}$$

has always a solution in $\mathring{W}_{2,1}^k(\varepsilon\Omega)$. Then Theorem 4.3.7 is proved since A is invariant under the transformation $x \to \varepsilon x$.

Extend f to a function $h \in \mathring{W}_2^k(\Omega)$. In virtue of (4.3.25) there exists some linear combination h_1 of the functions $f_1,\dots,f_q$ such that $h+h_1 \in A(\overset{\circ}{W}{}^k_{2,1}(\Omega))$. Thus the equation $Av = h+h_1$ has a solution $v \in \overset{\circ}{W}{}^k_{2,1}(\Omega)$ so that $u = v|_{\varepsilon\Omega} \in \overset{\circ}{W}{}^k_{2,1}(\varepsilon\Omega)$ is a solution of (4.3.26) Q.E.D. □

4.3.4. Finally, as an example, we consider the operator (4.o.1) which has already been studied in Sec. 4.2. By virtue of (4.2.1), the operator function (4.3.4) corresponding to (4.o.1) is

$$\mathcal{A}_z = (iz)^2 + (n-2+a)iz + b - \delta .$$

Using the fact that $m(m+n-2)$, $m \in \mathbb{N}_0$, are the eigenvalues of the Beltrami operator δ, we obtain that $\mathcal{A}_z$ has the eigenvalues

$$z_{jm} = \mu_j(m)/i \ , \ j=1,2, \ m \in \mathbb{N}_0$$

where $\mu_j(m)$ are the numbers defined by (4.2.11). Then Theorems 4.3.2, 4.3.5 and 4.3.7 yield the following results for the operator A defined in (4.o.1) which, in part, generalize those of Theorem 4.2.2:

(i) $A \in L(W^k_{2,2}(\mathbb{R}^n), W^k_2(\mathbb{R}^n))$ is an injective Fredholm operator if

$$\mathrm{Re}\,\mu_j(m) \neq -n/2, \ -n/2+k \ , \ j=1,2, \ m \in \mathbb{N}_0$$

which is invertible for k=0.

(ii) $A \in L(\overset{\circ}{W}{}^k_{2,2}(\Omega), \overset{\circ}{W}{}^k_2(\Omega))$ is surjective when Ω is bounded and

$$\mathrm{Re}\,\mu_j(m) \neq -n/2+k, \ j=1,2, \ m \in \mathbb{N}_0 \ .$$

4.4. A hypoelliptic differential operator which is not locally solvable

We consider the differential operator

$$A = r^{2k}(x_1\,\partial/\partial x_1 + x_2\,\partial/\partial x_2) + ir^{2k}(x_1\,\partial/\partial x_2 - x_2\,\partial/\partial x_1) + 1 \qquad (4.4.1)$$

in $\mathbb{R}^2$, where $r^2 = x_1^2 + x_2^2$, $i = \sqrt{-1}$ and $k \in \mathbb{N}$. Note that (4.4.1) is elliptic outside the origin. We say that a differential operator P with C^∞ coefficients is hypoelliptic at a point x if there exists an open set U containing x such that if $u \in \mathcal{D}'(U)$ and $Pu \in C^\infty(U)$, $u \in C^\infty(U)$. It is clear that (4.4.1) is hypoelliptic and locally solvable at every point $x \in \mathbb{R}^2 \setminus \{0\}$. The aim of this section is to prove

Theorem 4.4.1. The operator (4.4.1) is hypoelliptic but not locally solvable at the origin.

We first introduce some notation and prove two lemmas. The spherical (polar) coordinates in $\mathbb{R}^2$ are given by

$$x_1 = r\cos\varphi\ ,\ x_2 = r\sin\varphi\ ,\ \varphi\in[0,2\pi)$$

and relation (4.1.4) reads

$$\begin{aligned}\partial/\partial x_1 &= \cos\varphi\ \partial/\partial r - r^{-1}\sin\varphi\ \partial/\partial\varphi\ ,\\ \partial/\partial x_2 &= \sin\varphi\ \partial/\partial r + r^{-1}\cos\varphi\ \partial/\partial\varphi\ .\end{aligned} \tag{4.4.2}$$

Moreover, as a special case of (4.2.2) - (4.2.4), we can write

$$u(x) = \sum_{m\in\mathbb{Z}} u_m(r)e^{im\varphi},\ u_m(r) = (2\pi)^{-1}\int_0^{2\pi} u(x)e^{-im\varphi}\,d\varphi \tag{4.4.3}$$

for any $u\in L_2^{\varrho}(\Omega_R)$, $0<R<\infty$, $\varrho\in\mathbb{R}$, and

$$\|u\|^2_{L_2^{\varrho}(\Omega_R)} = \sum_{m\in\mathbb{Z}} \|u_m\|^2_{L_2^{\varrho-1/2}(0,R)}\ . \tag{4.4.4}$$

A and its transpose ${}^tA$ can be written in polar coordinates

$$A = r^{2k+1}\,\partial/\partial r + ir^{2k}\,\partial/\partial\varphi + 1$$

$${}^tA = -\,r^{2k+1}\,\partial/\partial r - ir^{2k}\,\partial/\partial\varphi + 1 - (2k+2)r^{2k}\ .$$

Furthermore, for any $u\in C^{\infty}(\bar{\Omega}_R)$, we can write

$$Au = \sum_{m\in\mathbb{Z}} (A_m u_m)e^{im\varphi}\ ,\ A_m = r^{2k+1}d/dr - 1 - mr^{2k}\ , \tag{4.4.5}$$

$${}^tAu = -\sum_{m\in\mathbb{Z}} (B_m u_m)e^{im\varphi}\ ,\ B_m = r^{2k+1}d/dr - 1 + (2k+2-m)r^{2k}\ . \tag{4.4.6}$$

Define the function

$$h_m(r) = (2k)^{-1}r^{-2k} + (2k+2-m)\ln r$$

on $(0,R]$ and set

$$d_m = (2k+2-m)^{-1/2k}\ (m\le 0),\ d_m = R\ (m>0)\ .$$

<u>Lemma 4.4.2.</u> For any $\varrho\in\mathbb{R}$ and $m\in\mathbb{Z}$, the equation

$$B_m v = g,\ g\in L_2^{\varrho+2k+1}(0,R) \tag{4.4.7}$$

has the solution

$$v(r) = B_m^{-1}g = e^{-h_m(r)}\int_{d_m}^{r} e^{h_m(s)}s^{-2k-1}g(s)\,ds \tag{4.4.8}$$

in $L_2^{\varrho}(0,R)$ which can be estimated by

$$\|v\|_{L_2^{\varrho}(0,R)} \le c\,\|g\|_{L_2^{\varrho+2k+1}(0,R)}\ , \tag{4.4.9}$$

where the constants $c = c(\rho)$ do not depend on m and g.

Proof. It follows from Theorem 2.3.2 that (4.4.8) belongs to $L_2^{\rho}(0,R)$ and satisfies equation (4.4.7) on (0,R). Estimate (2.3.12), however, does not suffice for our purpose since it is not uniform with respect to m. We now show, in three steps, that (4.4.8) satisfies (4.4.9).

1. Let $m > 0$. Then the function $h_{m\rho} = h_m + \rho \ln r$ is decreasing on (0,R) for all sufficiently large m. Putting $g_1 = r^{-2k-1-\rho}g$, we get

$$|r^{-\rho}v(r)| \leq e^{-h_{m\rho}(r)} \int_r^R e^{h_{m\rho}(s)} |g_1(s)|\,ds \leq \int_r^R |g_1(s)|\,ds$$

$$\leq R^{1/2} \|g\|_{L_2^{\rho+2k+1}(0,R)}, \quad r \in (0,R) \;,$$

hence (4.4.9) when m is large enough. By Theorem 2.3.2, (4.4.9) holds for a finite number of indices m so that the assertion is proved when $m > 0$.

2. Let $m \leq 0$ and $\rho \geq 0$. $h_{m\rho}$ is decreasing on $(0,d_{m\rho})$ and increasing on $(d_{m\rho},R)$, where $d_{m\rho} = (2k+2-m+\rho)^{-1/2k}$. Note that $d_m = d_{mo}$ and $h_m = h_{mo}$. For $r \in (d_m,R)$, we then have

$$|r^{-\rho}v(r)| \leq e^{-h_{m\rho}(r)} \int_{d_m}^r e^{h_{m\rho}(s)} |g_1(s)|\,ds \leq \int_{d_m}^r |g_1(s)|\,ds$$

$$\leq c_1 \|g\|_{L_2^{\rho+2k+1}(0,R)} \;,$$

and for $r \in (0,d_m)$,

$$|r^{-\rho}v(r)| \leq e^{-h_{m\rho}(r)} \Big\{ \int_{d_{m\rho}}^{d_m} e^{h_{m\rho}(s)} |g_1(s)|\,ds + \Big|\int_{d_{m\rho}}^r e^{h_{m\rho}(s)} |g_1|\,ds\Big| \Big\}$$

$$\leq e^{h_{m\rho}(d_m)-h_{m\rho}(d_{m\rho})} \int_{d_{m\rho}}^{d_m} |g_1|\,ds + \Big| \int_{d_{m\rho}}^r |g_1|\,ds \Big|$$

$$\leq (c_2(\rho)+1)R^{1/2} \|g\|_{L_2^{\rho+2k+1}(0,R)} \;,$$

$$c_2(\rho) = e^{-\rho/2k} \{(2k+2-m+\rho)/(2k+2-m)\}^{(2k+2-m+\rho)/2k} \;.$$

Thus we obtain (4.4.9) for $\rho \geq 0$.

3. Let $m \leq 0$ and $\rho < 0$. If $2k+2-m+\rho > 0$, similarly as in 2. we get

$$|r^{-\varrho} v(r)| \le c_1 \|g\|_{L_2^{\varrho+2k+1}(0,R)}$$

when $r \in (0,d_m)$ and

$$|r^{-\varrho} v(r)| \le e^{-h_{m\varrho}(r)} \Big\{ \int_{d_m}^{d_{m\varrho}} e^{h_{m\varrho}(s)} |g_1| ds + \Big| \int_{d_{m\varrho}}^{r} e^{h_{m\varrho}(s)} |g_1| ds \Big| \Big\}$$

$$\le c_2(\varrho) \|g\|_{L_2^{\varrho+2k+1}(0,R)}$$

when $r \in (d_m,R)$, hence (4.4.9). Since estimate (4.4.9) for $m \ge 2k+2+\varrho$ follows from Theorem 2.3.2 again, the lemma is proved. □

Next, we prove that ${}^tA$ is surjective in the space of flat C^∞ functions. Consider the equation

$$^tAu = f, \; f = \sum_{m\in\mathbb{Z}} f_m(r)e^{im\varphi} \in C^{\infty,o}(\bar{\Omega}_R) \; . \qquad (4.4.1o)$$

<u>Lemma 4.4.3.</u> For every right-hand side, (4.4.1o) has the solution

$$u = {}^tA^{-1}f = -\sum_{m\in\mathbb{Z}} (B_m^{-1}f_m)e^{im\varphi} \qquad (4.4.11)$$

in $C^{\infty,o}(\bar{\Omega}_R)$ with B_m^{-1} defined in (4.4.8).

<u>Proof.</u> Set $u_m = B_m^{-1}f_m$. From Lemma 4.4.2 we obtain the estimates

$$\sum_{m\in\mathbb{Z}} m^{2n} \|u_m\|^2_{L_2^{\varrho-1/2}(0,R)} \le c(\varrho) \sum_{m\in\mathbb{Z}} m^{2n} \|f_m\|_{L_2^{\varrho+2k+1/2}(0,R)} \qquad (4.4.12)$$

for any $n \in \mathbb{N}_o$ and $\varrho \in \mathbb{R}$. With u defined by (4.4.11), (4.4.12) and (4.4.4) imply

$$(\partial/\partial\varphi)^n u \in L_2^{\varrho}(\Omega_R) \; , \quad \varrho \in \mathbb{R}, \; n \in \mathbb{N}_o \; . \qquad (4.4.13)$$

Furthermore, (4.4.11) is a solution of equation (4.4.1o). Indeed, since $u_m, du_m/dr \in L_2^{\varrho}(0,R)$ for any ϱ and m, (4.4.11) and (4.4.5) imply

$$\langle f,v \rangle = -\sum_{m\in\mathbb{Z}} \int_o^R (B_m u_m) v_m r dr = \sum_{m\in\mathbb{Z}} \int_o^R u_m(A_m v_m) r dr$$

$$= \langle u,Av \rangle \; , \; v \in C_o^\infty(\Omega_R) \; ,$$

where v_m are the coefficients of the spherical harmonics expansion of v. (4.4.13) with n = 1 and ${}^tAu \in C^{\infty,o}(\bar{\Omega}_R)$ now imply $r^{2k+1}\partial u/\partial r \in L_2(\Omega_R)$ for any ϱ . Analogously, (4.4.13) and

$$ {}^tA(\partial/\partial\varphi)^n u = (\partial/\partial\varphi)^n\, {}^tAu \in C^{\infty,0}(\bar\Omega_R) $$

imply $r^{2k+1}\partial/\partial r(\partial/\partial\varphi)^n u \in L_2^\varrho(\Omega_R)$ for all ϱ and $n \geq 2$, and applying the operator $r^{2k+1}\partial/\partial r$ to both sides of equation (4.4.1o), we successively obtain

$$ (r^{2k+1}\partial/\partial r)^m(\partial/\partial\varphi)^n u \in L_2^\varrho(\Omega_R),\ m,n \in \mathbb{N}_0,\ \ \varrho \in \mathbb{R}\ . \qquad (4.4.14) $$

Finally, we verify that (4.4.14) implies $u \in C^{\infty,0}(\bar\Omega_R)$. Indeed, by (4.4.14) and (4.4.2), the distribution u in $\Omega_R \setminus \{0\}$ coincides with some function $\tilde u$ satisfying $D_x^\alpha \tilde u \in L_2^\varrho(\Omega_R)$ for all ϱ and all multi-indices α. Hence $\tilde u \in C^{\infty,0}(\bar\Omega_R)$, and $u = \tilde u$ in Ω_R because $u \in L_2(\Omega_R)$. □

Now we are ready to prove Theorem 4.4.1.

<u>Proof of hypoellipticity.</u> We first observe that it is sufficient to show the implication $u \in \mathcal{E}'(\Omega_R)$, $Au \in C^{\infty,0}(\bar\Omega_R) \Rightarrow u \in C^\infty(\bar\Omega_R)$ for some $R > 0$, where $\mathcal{E}'(\Omega_R)$ denotes the space of distributions with compact support in Ω_R:

Let $u \in \mathcal{D}'(\Omega_R)$ and $Au \in C^\infty(\Omega_R)$. Set $v = \chi u$, where $\chi \in C_0^\infty(\Omega_R)$ and $\chi = 1$ in $\Omega_{R/2}$. By the ellipticity of A in $\mathbb{R}^2 \setminus \{0\}$, $Av \in C^\infty(\bar\Omega_R)$, and $u \in C^\infty(\Omega_R)$ if and only if $v \in C^\infty(\bar\Omega_R)$. Furthermore, it is easily seen that there exists a formal power series p such that $Ap = \hat v$, where the hat denotes the formal Taylor series of the corresponding C^∞ function at the origin. By a theorem of Borel (cf. Narasimhan [1, Chap. 1]), there is a function $v_1 \in C^\infty(\bar\Omega_R)$ satisfying $\hat v_1 = p$. Thus $Aw \in C^{\infty,0}(\bar\Omega_R)$ with $w = v - \chi v_1 \in \mathcal{E}'(\Omega_R)$, and we have $v \in C^\infty(\bar\Omega_R)$ if and only if $w \in C^\infty(\bar\Omega_R)$.

Let now $u \in \mathcal{E}'(\Omega_R)$, $Au \in C^{\infty,0}(\bar\Omega_R)$ and $\psi \in C^{\infty,0}(\bar\Omega_R)$. Then

$$ \langle u,\psi\rangle = \langle u, {}^tA\,{}^tA^{-1}\psi\rangle = \langle Au, {}^tA^{-1}\psi\rangle $$

with ${}^tA^{-1}$ defined in (4.4.11). Applying estimate (4.4.12) for $n = 0$, we obtain

$$ |\langle u,\psi\rangle| \leq \|Au\|_{L_2^{\varrho+2k+1}(\Omega_R)} \|{}^tA^{-1}\psi\|_{L_2^{-\varrho-2k-1}(\Omega_R)} $$

$$ \leq c(\varrho)\|\psi\|_{L_2^{-\varrho}(\Omega_R)}\ ,\quad \psi \in C^{\infty,0}(\bar\Omega_R)\ . $$

Since $C^{\infty,0}$ is dense in $L_2^{-\varrho}$ and $L_2^{-\varrho}$ is the dual space of L_2^{ϱ} with respect to the bilinear form $\langle\cdot,\cdot\rangle$, the last inequality yields $u \in L_2^{\varrho}(\Omega_R)$ for any ϱ. Moreover, because of

$$A(\partial/\partial\varphi)^n u \in C^{\infty,0}(\bar{\Omega}_R), \quad (\partial/\partial\varphi)^n u \in \mathcal{E}'(\Omega_R), \quad n \in \mathbb{N},$$

it follows in the same way that $(\partial/\partial\varphi)^n u \in L_2^{\varrho}(\Omega_R)$, $\varrho \in \mathbb{R}$. Finally, as in the proof of Lemma 4.4.3 we get (4.4.14), hence $u \in C^{\infty,0}(\bar{\Omega}_R)$. □

Proof of nonsolvability. Choose $\chi \in C_0^{\infty}(0,R)$ such that $0 \le \chi \le 1$ on $(0,R)$ and $\chi = 1$ on $[R/3, R/2]$. We set

$$f = \sum_{m\in\mathbb{Z}} f_m(r)e^{im\varphi}, \quad f_m = \alpha^{-m}\chi \ (m \ge 0), \quad f_m = 0 \ (m < 0),$$

where $\alpha > 1$ will suitable be chosen later on. As in the proof of Theorem 4.2.3 one can verify that $f \in C_0^{\infty}(\Omega_{2R})$. Suppose the equation $Au = f$ has a solution $u \in \mathcal{D}'(\Omega_{2R})$. Since A is hypoelliptic, $u \in C^{\infty}(\bar{\Omega}_R)$. Therefore, by (4.4.3) and (4.4.5), $u = \sum u_m e^{im\varphi}$ with $u_m \in C^{\infty}[0,R]$ and $A_m u_m = f_m$. By Theorem 2.3.2, the latter equation has the unique solution

$$u_m(r) = \int_0^r \exp\{(r^{-2k}-s^{-2k})/2k\}\, s^{-2k-1}(r/s)^m f_m(s)ds$$

in $C^{\infty}[0,R]$. Thus, for all $r \in [2R/3,R]$ and $m \ge 0$,

$$|u_m(r)| \ge \int_{R/3}^{R/2} \exp\{(r^{-2k}-s^{-2k})/2k\}\, s^{-2k-1}(4/3)^m \alpha^{-m} ds \ge c(4/3\alpha)^m.$$

Choosing $\alpha \in (1,4/3)$, by (4.4.4) one obtains

$$\|u\|^2_{0,\Omega_R} = \sum_{m\ge 0} \|u_m\|^2_{L_2^{-1/2}(0,R)} \ge \sum_{m\ge 0} \|u_m\|^2_{L_2^{-1/2}(2R/3,R)}$$

$$\ge c \sum_{m\ge 0} (4/3\alpha)^m = \infty,$$

hence a contradiction. This finishes the proof of the theorem. □

Remark 4.4.4. One can prove the following stronger nonsolvability result for the operator (4.4.1) (see Lorenz [1]):
There exists $g \in C_0^{\infty}(\Omega_R)$ such that $A(\mathcal{D}'(\Omega_r)) \not\ni g$ for any $r \in (0,R)$.

4.5. Comments and references

4.2. Baouendi, Goulaouic and Lipkin [1] studied the operator (4.0.1) in the space of germs of analytic functions at the origin and gave a complete description of the kernel and the range of this operator. Theorems 4.2.2 and 4.2.4 were proved in Elschner and Lorenz [3]. As a special case of Theorem 4.2.3, the nonsolvability in flat functions for the operator $r^2\Delta$ in the plane was shown by Alinhac and Baouendi [1]. In Elschner and Lorenz [3] one can find further results on the kernel and the range of (4.o.1) in the Sobolev spaces $W_2^{2k}(\Omega_R)$, $k \in \mathbb{N}$.

4.3. The class of operators considered here was introduced by Baouendi and Sjöstrand [1], [2] who studied analytic regularity and hypoellipticity. In particular, they proved in [1] that the operator (4.3.1) is not hypoelliptic at the origin if (4.3.2) and a certain additional condition are satisfied. The results of this section are due to Elschner and Lorenz [1], [2]. Our methods are similar to those of Kondratiev [1] and Bagirov and Kondratiev [1] who studied elliptic differential operators in domains with corners or angular points and in $\mathbb{R}^n$. Lewis and Parenti [1] used Mellin transform techniques to study the solvability properties of singular abstract evolution equations of the form

$$tu'(t) + Au(t) = f(t)$$

where A is a closed densely defined operator on a Hilbert space.

4.4. Here we followed Elschner and Lorenz [4]. An example of an unsolvable hypoelliptic equation was first given by Kannai [1]. His operator is defined in $\mathbb{R}^2$ and degenerates on a line. Note that (4.4.1) is not contained in the classes of unsolvable operators considered in Hörmander [2] and Beals [1]. The method of this section canbe extended to certain differential operators of higher order; see Lorenz [1], [2]. However, up to now there are no general and complete results on hypoellipticity and local solvability for a wider class of elliptic operators degenerating at one point different from that in 4.3.

5. DEGENERATE PSEUDODIFFERENTIAL OPERATORS ON A CLOSED CURVE

In this chapter we consider classical pseudodifferential operators (PDOs) on a simple closed C^{∞} curve in the complex plane. This class of operators contains singular integral and integro-differential operators and certain integral operators with logarithmic kernels as special cases. Following Agranovič, we first give a global definition of PDOs on a closed curve, using the Fourier series of periodic functions. Then we collect some of their basic properties including the well-known elliptic a priori estimates and Gårding-Melin inequalities, which allow an elementary proof in the one-dimensional case.

Using a priori estimates based on Melin's inequality, we study Fredholm property and index in Sobolev spaces and in C^{∞} for a class of PDOs whose principal symbol may have rather general degeneracies and whose subprincipal symbol satisfies a certain condition on the characteristic set. The index will be expressed as a winding number by means of the principal and the subprincipal symbol of the operator. Furthermore, using a classical reduction to the boundary, we derive an index formula as well as an existence and uniqueness theorem for the degenerate oblique derivative problem in the plane. Finally, more general results on the index of degenerate PDOs are obtained when the principal symbol has only zeros of finite order. The index then may depend on an arbitrary (finite) number of terms of the complete symbol.

5.1. Pseudodifferential operators on a closed curve

5.1.1. Let Γ be a simple closed curve in the complex plane which is the positively oriented boundary of a bounded domain Ω and has the equation $t = t(x)$, $x \in [-\pi, \pi]$. We assume that Γ is a C^{∞} curve, i.e. $t(x)$ belongs to the space C^{∞} of all 2π-periodic infinitely differentiable (complex-valued) functions on the real axis. In the following we often identify functions $u(t)$ defined on Γ with 2π-periodic functions on $\mathbb{R}$:

$$u(x) = (u \circ t)(x) = u(t(x)) \ . \qquad (5.1.1)$$

By W_2^k we denote the periodic Sobolev space of order $k \in \mathbb{R}$, i.e. the completion of C^∞ with respect to the norm

$$\|u\|_k = \{ |\hat{u}_o|^2 + \sum_{0\neq m\in\mathbb{Z}} |\hat{u}_m|^2 |m|^{2k} \}^{1/2} , \qquad (5.1.2)$$

where

$$\hat{u}_m = (2\pi)^{-1/2} \int_{-\pi}^{\pi} u(x) e^{-imx} dx$$

are the Fourier coefficients of u. W_2^k is of course a Hilbert space with the scalar product

$$(u,v)_k = \hat{u}_o \bar{\hat{v}}_o + \sum_{0\neq m\in\mathbb{Z}} \hat{u}_m \bar{\hat{v}}_m |m|^{2k} , \qquad (5.1.3)$$

where $\hat{v}_m$ are the Fourier coefficients of v. For $L_2 = W_2^o$, we have

$$(u,v)_o = \sum_{m\in\mathbb{Z}} \hat{u}_m \bar{\hat{v}}_m = \int_{-\pi}^{\pi} u(x)\overline{v(x)} dx . \qquad (5.1.4)$$

It is well-known that, for $k \geq 0$,

$$\|u\|_k^2 = \|u\|_o^2 + \|D^k u\|_o^2 \quad \text{if } k \in \mathbb{N} ,$$

$$\|u\|_k^2 = \|u\|_o^2 + \iint_{-\pi}^{\pi} |D^n u(x) - D^n u(y)|^2 |x-y|^{-1-2\delta} dxdy \qquad (5.1.5)$$

$$\text{if } k = n+\delta,\ n \in \mathbb{N}_o,\ \delta \in (0,1)$$

is equivalent to the norm (5.1.2) (cf. Lions and Magenes [1, Chap.1.7]). Here $D = D_x$ stands for the differentiation d/dx. For any $j,k \in \mathbb{R}$, W_2^{j-k} can be identified with the antidual of W_2^{j+k} with respect to the Hermitean form $(u,v)_j$ extended to $u \in W_2^{j-k}$ and $v \in W_2^{j+k}$, and

$$\|u\|_{j-k} = \sup_{v\in W_2^{j+k},\, v\neq 0} |(u,v)_j| / \|v\|_{j+k} . \qquad (5.1.6)$$

Note that the Fréchet space C^∞ is the projective limit of the spaces W_2^k $(k \in \mathbb{R})$, while the space $\mathcal{D}'$ of 2π-periodic distributions on $\mathbb{R}$ is the inductive limit of the W_2^k.

Let us define the operator

$$\Lambda u = (2\pi)^{-1/2} \{ \hat{u}_o + \sum_{0\neq m\in\mathbb{Z}} \hat{u}_m |m| e^{imx} \} , \ u \in C^\infty .$$

Its closure in L_2 is a positive definite selfadjoint operator with the domain of definition $D(\Lambda) = W_2^1$, which will also be denoted by Λ. It has the eigenvalues $\lambda_o = 1$, $\lambda_m = |m|$ $(m \in \mathbb{Z} \setminus \{0\})$ which correspond to the orthonormal system of eigenfunctions $\{ e^{imx}/\sqrt{2\pi}, m \in \mathbb{Z} \}$. Hence, for all $k \in \mathbb{R}$, the operators Λ^k are defined by

$$\Lambda^k u = (2\pi)^{-1/2}\{\hat{u}_o + \sum_{0\neq m\in\mathbb{Z}} \hat{u}_m |m|^k e^{imx}\},\ D(\Lambda^k) = W_2^k, \tag{5.1.7}$$

and we have $(u,v)_k = (\Lambda^k u, \Lambda^k v)_o$ for all $k \in \mathbb{R}$ and $u,v \in W_2^k$. Therefore $\{W_2^k, k \in \mathbb{R}\}$ is a Hilbert scale, and for any linear operator A satisfying

$$\|A\|_{L(W_2^{k_j}, W_2^{l_j})} \leq c_j \quad (j=1,2),\ k_1 < k_2,\ l_1 < l_2,$$

$$k = (1-\theta)k_1 + \theta k_2,\ l = (1-\theta)l_1 + l_2\theta,\ \theta \in (0,1),$$

one obtains the estimate (see Triebel [4, Chap. 1.18.1o])

$$\|A\|_{L(W_2^k, W_2^l)} \leq c_1^{1-\theta} c_2^{\theta}. \tag{5.1.8}$$

Finally, by $W_2^k(\Gamma)$ we denote the closure of $C^\infty(\Gamma)$ with respect to the norm $\|u \circ t\|_k$. The spaces $W_2^k(\Gamma)$ and $C^\infty(\Gamma)$ may be identified with W_2^k and C^∞, respectively, in a natural way.

5.1.2. For a function $a(t,\xi)$ defined on a subset of $\Gamma \times \mathbb{R}$, set $a(x,\xi) = a(t(x),\xi)$, and let $\psi \in C^\infty(\mathbb{R})$ be some excision function satisfying $\psi(\xi) = 0$ when $|\xi| \leq 1/2$ and $\psi(\xi) = 1$ when $|\xi| \geq 1$.

Definition 5.1.1. Let $l \in \mathbb{R}$. We define the class CS^l to consist of the set of $a \in C^\infty(\Gamma \times \mathbb{R})$ with the property that there exist functions

$$a_{l-j} \in C^\infty(\Gamma \times (\mathbb{R} \setminus \{0\})),\ j \in \mathbb{N}_o$$

which are positive homogeneous of degree $l-j$ in ξ and constants $c_{N\alpha\beta}$ such that

$$\begin{aligned} &|D_x^\alpha D_\xi^\beta [a(x,\xi) - \sum_{0\leq j\leq N} \psi(\xi) a_{l-j}(x,\xi)]| \\ &\leq c_{N\alpha\beta}\,(1+|\xi|)^{l-N-1-\beta},\ (x,\xi) \in \mathbb{R}\times\mathbb{R} \end{aligned} \tag{5.1.9}$$

for any $\alpha, \beta, N \in \mathbb{N}_o$. Instead of (5.1.9), we also write $a \sim a_l + a_{l-1} + \dots$. Note that the class CS^l does not depend on the choice of the excision function ψ. Furthermore, for any $j \in \mathbb{N}_o$, we obviously have the representation

$$a_{l-j}(t,\xi) = [a_{l-j}^+(t)h(\xi) + a_{l-j}^-(t)h(-\xi)]\ |\xi|^{l-j}, \tag{5.1.1o}$$

where

$$a^{\pm}_{1-j}(t,\xi) = a_{1-j}(t,\pm 1) \in C^{\infty}(\Gamma),\ h(\xi) = 2^{-1}(1+\xi/|\xi|)\ .$$

Definition 5.1.2. An integral operator of the form

$$(Ku)(x) = \int_{\Gamma} k(t,\tau)u(\tau)d\tau = \int_{-\pi}^{\pi} k(t(x),t(y))u(t(y))t'(y)dy,$$

$$k \in C^{\infty}(\Gamma \times \Gamma)$$

is called a smoothing operator.

Definition 5.1.3. Let $a \in CS^{1}$ and $a \sim a_{1} + a_{1-1} + \ldots$. A is called a classical pseudodifferential operator (PDO) of order 1 on Γ with (complete) symbol $\sigma_{A} = a_{1} + a_{1-1} + \ldots$ if

$$Au = \sum_{m \in \mathbb{Z}} (2\pi)^{-1/2} a(x,m)\hat{u}_{m} e^{imx} + Ku,\ u \in C^{\infty}(\Gamma)\ , \tag{5.1.11}$$

where $\hat{u}_{m}$ are the Fourier coefficients of $u \circ t$ and K is some smoothing operator. In this case we also say that A belongs to the class $OPCS^{1}$. The term a_{1} in σ_{A} is called the principal symbol of A.

Before studying the basic properties of PDOs, we consider some examples.

Example 5.1.4. Consider the Cauchy integral operator

$$(S_{\Gamma}u)(t) = (\pi i)^{-1}\int_{\Gamma} u(\tau)(\tau - t)^{-1}d\tau$$

on Γ, where the integral is to be interpreted as a Cauchy principal value (see Muschelischwili [1]). We show that S_{Γ} is a PDO of order 0 with symbol $\xi/|\xi|$. This is easy to check in the case of the unit circle $\Gamma_{0} = \{z = e^{ix} : x \in [-\pi,\pi]\}$. Indeed, if $u \in C^{\infty}(\Gamma_{0})$ and $\hat{u}_{m}$ are the Fourier coefficients of $u(e^{ix})$, we obtain

$$(2\pi)^{1/2} S_{\Gamma_{0}} u = \sum_{m \geq 0} \hat{u}_{m} z^{m} - \sum_{m < 0} \hat{u}_{m} z^{m} = \sum_{m \in \mathbb{Z}} a_{0}(m)\hat{u}_{m} e^{imx} + K_{0}u\ ,$$

$$a_{0}(\xi) = \psi(\xi)\,\xi/|\xi|\ ,\ K_{0}u = \int_{-\pi}^{\pi} u dx\ . \tag{5.1.12}$$

In the general case there exists a conformal mapping β of the unit disk onto Ω which is a C^{∞} diffeomorphism of Γ_{0} onto Γ (cf. Vekua [1, Th. 1.9]). Then we get

$$S_{\Gamma} u = (\pi i)^{-1} \int_{\Gamma_{0}} u(\beta(\zeta))[\beta(\zeta) - \beta(z)]^{-1} \beta'(\zeta)d\zeta$$

$$= S_{\Gamma_{0}}(u \circ \beta) + K_{1}(u \circ \beta),\ u \in C^{\infty}(\Gamma),\ \beta(z) \in \Gamma\ ,$$

where

$$(K_1u)(z) = \int_{\Gamma_0} k_1(z,\zeta)u(\zeta)d\zeta \ , \ z \in \Gamma_0 \ ,$$

$$k_1(z,\zeta) = (\zeta - z)^{-1} - \beta'(\zeta)[\beta(\zeta) - \beta(z)]^{-1} \in C^{\infty}(\Gamma_0 \times \Gamma_0) \ .$$

Consequently, by (5.1.12) we have

$$(2\pi)^{1/2}S_{\Gamma}u = \sum_{m\in\mathbb{Z}} a_0(m)\hat{u}_m e^{imx} + Ku \ , \ u \in C^{\infty}(\Gamma) \ , \qquad (5.1.13)$$

where $\hat{u}_m$ are the Fourier coefficients of $u \circ t$ and K is a smoothing operator Q.E.D.

If $a,b \in C^{\infty}(\Gamma)$ and K is a smoothing operator, then

$$A = aI + bS_{\Gamma} + K$$

is called a singular integral operator on Γ. It has the symbol $a+b\,\xi/|\xi|$.

Example 5.1.5. The differential operator

$$A = \sum_{0\le j\le m} a_j(x)D^j \ , \ a_j \in C^{\infty}$$

is a PDO of order m on Γ with symbol $a_0(x)+a_1(x)i\xi+\ldots+a_m(x)(i\xi)^m$.

Example 5.1.6. The operator Λ^k defined by (5.1.7) is a PDO of order k on Γ with symbol $|\xi|^k$.

5.1.3. We now sketch the proofs of some important properties of PDOs which we need in the sequel. Define the operators

$$P^+u = \sum_{m\ge 0}(2\pi)^{-1/2}\hat{u}_m e^{imx} \ , \ P^-u = \sum_{m<0}(2\pi)^{-1/2}\hat{u}_m e^{imx}, u \in C^{\infty} \qquad (5.1.14)$$

which are PDOs of order 0 on Γ with symbols $h(\xi)$, $h(-\xi)$.

Theorem 5.1.7. For any $N \in \mathbb{N}_0$, the PDO (5.1.11) can be written

$$A = \sum_{0\le j\le N}[a^+_{l-j}P^+ + a^-_{l-j}P^-]\Lambda^{l-j} + A_N \ , \qquad (5.1.15)$$

where $A_N \in OPCS^{l-N-1}$ and $a^{\pm}_{l-j}$ are defined in (5.1.10).

Proof. Setting

$$b_N = \sum_{0\le j\le N}\gamma(\xi)[a^+_{l-j}h(\xi) + a^-_{l-j}h(-\xi)]\ |\xi|^{l-j} \ ,$$

by definitions (5.1.7) and (5.1.14) we obtain

$$(2\pi)^{1/2}\sum_{0\le j\le N}[a^+_{l-j}P^+ + a^-_{l-j}P^-]\Lambda^{l-j}u = \sum_{m\in\mathbb{Z}} b_N(x,m)\hat{u}_m e^{imx} + K_0u$$

with $u \in C^{\infty}$ and K_0 defined in (5.1.12). Since $a-b_N \in CS^{l-N-1}$ in view of

(5.1.9), (5.1.15) is proved. □

It is easy to see that (5.1.11) is a continuous map of C^∞ into itself. We now investigate its continuity in Sobolev spaces.

<u>Theorem 5.1.8.</u> For any $k \in \mathbb{R}$, the PDO (5.1.11) can be extended to an operator $A \in L(W_2^k, W_2^{k-1})$.

<u>Proof.</u> We first consider a smoothing operator K with kernel $k(x,y)$:

$$Ku = \int_{-\pi}^{\pi} k(x,y)u(y)dy \;.$$

Since $D^k K$ is an integral operator with kernel $D_x^k k$, we obviously have $K \in L(W_2^0, W_2^k)$ for any $k \in \mathbb{N}_0$. Let K^* be the integral operator with kernel $\overline{k(y,x)}$. Using the relations

$$(Ku,v)_0 = (u,K^*v)_0 \;,\; u,v \in C^\infty$$

and (5.1.6), one obtains $K \in L(W_2^{-k}, W_2^0)$ for all $k \in \mathbb{N}_0$. Thus $K \in L(W_2^k, W_2^{k-l})$ for all $k,l \in \mathbb{Z}$, and by interpolation (see (5.1.8)), this is valid for all $k,l \in \mathbb{R}$.

Consider now representation (5.1.15). It is sufficient to prove that $A_N \in L(W_2^k, W_2^{k-1})$ for some N, since $P^{\pm} \in L(W_2^k)$, $\Lambda^{1-j} \in L(W_2^k, W_2^{k-1+j})$ and $aI \in L(W_2^k)$ for any $k \in \mathbb{R}$ and $a \in C^\infty$. Further, we may assume that

$$A_N u = \sum_{m \neq 0} a_N(x,m)\hat{u}_m e^{imx}, \; a_N \in CS^{1-N-1} \;.$$

Choose a natural number $\bar{r} \geq k-1$. By (5.1.9) and the Cauchy-Schwarz inequality, we then obtain

$$|D^r A_N u(x)|^2 \leq c(\sum_m |\hat{u}_m| \, |m|^{1-N-1+r})^2 \leq c_1 \sum_m |\hat{u}_m|^2 |m|^{2k} \;,$$

$$u \in C^\infty, \; x \in \mathbb{R}, \; r = 0,1,\dots,\bar{r}$$

when N is sufficiently large. Consequently,

$$\| A_N u \|_r^2 \leq c_1 \| u \|_k^2 \;,\; u \in C^\infty$$

so that $A_N \in L(W_2^k, W_2^{k-1})$ for suitable N Q.E.D. □

<u>Remark 5.1.9.</u> A map $K \in L(C^\infty)$ has continuous extensions $K \in L(W_2^k, W_2^{k-l})$ for all $k,l \in \mathbb{R}$ if and only if K is a smoothing operator. The sufficiency was proved above, while the necessity is a consequence of the Schwartz kernel theorem (cf. e.g. Gelfand and Vilenkin [1]).

Theorem 5.1.10. (i) For any formal series $\sigma = a_l + a_{l-1} + \ldots$ with a_{l-j} given by (5.1.10), there exists a PDO $A \in OPCS^l$ with symbol σ.
(ii) If two operators $A_1, A_2 \in OPCS^l$ have the same symbol, then $A_1 - A_2$ is a smoothing operator.

Proof. (i) Set

$$a(x,\xi) = \sum_{j \geq 0} \psi(\varepsilon_j \xi) a_{l-j}(x,\xi)$$

and choose $\varepsilon_j \in (0,1)$ so small that

$$|D_x^\alpha D_\xi^\beta \psi(\varepsilon_j \xi) a_{l-j}(x,\xi)| \leq 2^{-j}(1+|\xi|)^{l-j+1-\beta}, \quad (x,\xi) \in \mathbb{R} \times \mathbb{R}$$

for all $\alpha + \beta \leq j$. Then one can verify that the series converges in $C^\infty(\Gamma \times \mathbb{R})$ and that a satisfies (5.1.9); see Hörmander [3], Šubin [1].
(ii) If the symbols of A_1 and A_2 coincide, then $A = A_1 - A_2$ takes the form (5.1.11), where $a \in CS^l$ for all $l \in \mathbb{R}$. Therefore, by Theorem 5.1.8, $A \in L(W_2^k, W_2^{k-l})$ for any $k,l \in \mathbb{R}$ so that A is a smoothing operator in view of Remark 5.1.9. □

In the sequel we shall write $A_1 \cong A_2$ if $A_1 - A_2$ is a smoothing operator, and $A = Op(\sigma)$ for a PDO A with symbol σ.

Theorem 5.1.11. Let $A \in OPCS^l$ and $B \in OPCS^m$ with symbols σ_A and σ_B, respectively. Then AB belongs to $OPCS^{l+m}$ and has the symbol

$$\sigma_{AB} = \sigma_A \circ \sigma_B := \sum_{r \geq 0} (D_\xi^r \sigma_A)(i^{-1}D_x)^r \sigma_B / r! \tag{5.1.16}$$

Proof. We restrict ourselves to the case $l,m \in \mathbb{Z}$. If A or B is a smoothing operator, then by Theorem 5.1.8 and Remark 5.1.9 AB is smoothing so that the theorem is valid in this case. Suppose now that $A = aP^{\pm}\Lambda^l$ and $B = bP^{\pm}\Lambda^m$, where $a,b \in C^\infty$. The commutator $[aI, S_\Gamma] = aS_\Gamma - S_\Gamma aI$ takes the form

$$[aI, S_\Gamma]\, u = (\pi i)^{-1} \int_\Gamma [a(t) - a(\tau)] u(\tau)(\tau - t)^{-1} d\tau, \quad u \in C^\infty,$$

hence it is a smoothing operator. Since further $[\Lambda^l, P^{\pm}] = 0$ and $P^{\pm} \cong (I \pm S_\Gamma)/2$ (see (5.1.13)), we observe that $a P^{\pm}\Lambda^l b P^{\mp}\Lambda^m \cong 0$ and formula (5.1.16) holds in this case. Moreover, (5.1.16) is obvious when $A = aI$ or $B = P^{\pm}\Lambda^m$. So it remains to consider the case $A = P^{\pm}\Lambda^l$, $B = bI$. Let first $l \in \mathbb{N}_0$. Since then $AB \cong (\pm D/i)^l b P^{\pm}$, (5.1.16) follows immediately from Leibniz' rule. Let now $l = -1,-2,\ldots$. Consider the PDO

$Op(|\xi|^l \circ b)$ which exists by Theorem 5.1.10 (i). Then it is not difficult to check that $|\xi|^{-l} \circ (|\xi|^l \circ b) = b$, hence $\Lambda^{-l} Op(|\xi|^l \circ b) \equiv bI$ in view of Theorem 5.1.10 (ii) and the things just proved. Consequently, $Op(|\xi|^l \circ b) \equiv \Lambda^l bI$. This implies (5.1.16) again since $AB = P^{\pm}\Lambda^l bI \equiv \Lambda^l bP^{\pm}$.

Using Theorem 5.1.7 and the above special cases of (5.1.16), we see that $T = AB - Op(\sigma_A \circ \sigma_B) \in OPCS^l$ for any $l \in \mathbb{R}$. Therefore, by Theorem 5.1.8 and Remark 5.1.9, T is a smoothing operator which proves (5.1.16) in the general case. □

We define the formal adjoint A^* of a PDO A by

$$(Au,v)_0 = (u,A^*v)_0, \quad u,v \in C^\infty.$$

__Theorem 5.1.12.__ Let $A \in OPCS^l$ with symbol σ_A. Then $A^* \in OPCS^l$ with symbol

$$\sigma_{A^*} = \sum_{r \geq 0} D^r (i^{-1} D_x)^r \overline{\sigma}_A / r! \tag{5.1.17}$$

__Proof.__ Similarly as in Theorem 5.1.11 it suffices to verify the assertion for a PDO of the form

$$A = (a^+P^+ + a^-P^-)\Lambda^j, \quad a^{\pm} \in C^\infty, \quad j \in \mathbb{R}.$$

Since $(P^{\pm})^* = P^{\pm}$, $(\Lambda^j)^* = \Lambda^j$ and $(a^{\pm}I)^* = \overline{a^{\pm}}I$, we have

$$A^* = \Lambda^j(P^+\overline{a^+}I + P^-\overline{a^-}I)$$

and (5.1.17) is a consequence of (5.1.16). □

__Definition 5.1.13.__ $A \in OPCS^l$ is called elliptic if its principal symbol satisfies $a_l(t,\xi) \neq 0$ for all $t \in \Gamma$ and $\xi = \pm 1$.

__Theorem 5.1.14.__ If $A \in OPCS^l$ is elliptic, then for any $k \in \mathbb{R}$ $A \in L(W_2^{k+l}, W_2^k)$ is a Fredholm operator with index

$$\text{ind } A = (2\pi)^{-1} \{\arg a_l(t,-1)/a_l(t,1)\}_\Gamma , \tag{5.1.18}$$

where $\{\cdot\}_\Gamma$ means the variation of the argument when t varies around the contour Γ in the positive direction. If A is not elliptic (i.e. degenerate), then $A \in L(W_2^{k+l}, W_2^k)$ is not Fredholm.

__Proof.__ In order to reduce the assertion to the case $k = l = 0$, we consider the operator $B = \Lambda^k A \Lambda^{-k-l}$ and the commutative diagram

$$\begin{array}{ccccc} A & : & W_2^{k+1} & \longrightarrow & W_2^k \\ & & \downarrow \Lambda^{k+1} & & \downarrow \Lambda^k \\ B & : & W_2^0 & \longrightarrow & W_2^0 \end{array} \quad . \tag{5.1.19}$$

By Theorems 5.1.7 and 5.1.11, we have

$$B = aI+bS_\Gamma +B_{-1},\ a = a_1^+ + a_1^-,\ b = a_1^+ - a_1^-,\ B_{-1} \in OPCS^{-1} .$$

Since the embedding $W_2^0 \subset W_2^{-1}$ is compact, Theorem 5.1.8 yields that the operator B_{-1} is compact in W_2^0. It is well-known that the singular integral operator $aI+bS_\Gamma$ is a Fredholm operator in W_2^0 with index (5.1.18) if $a_1^\pm(t) \neq 0$ for all $t \in \Gamma$ and that it is not Fredholm there if one of the functions $a_1^\pm$ vanishes at some point on Γ (see e.g. Prößdorf [1]). Since, by definition, the operators Λ^k and Λ^{k+1} are isomorphisms, the theorem follows from (5.1.19). □

<u>Corollary 5.1.15.</u> (elliptic regularity) Let $A \in OPCS^1$ be elliptic. Then for any $k_1, k_2 \in \mathbb{R}$, $k_1 < k_2$, $u \in W_2^{k_1+1}$ and $Au \in W_2^{k_2}$ imply $u \in W_2^{k_2+1}$. Moreover, if $u \in \mathcal{D}'$ and $Au \in C^\infty$, then $u \in C^\infty$.

<u>Proof.</u> Since $W_2^{k_2}$ is dense in $W_2^{k_1}$, the first assertion is a consequence of Theorem 5.1.14; see the proof of Theorem 1.5.1. The second assertion is then clear because of $C^\infty = \bigcap_k W_2^k$, $\mathcal{D}' = \bigcap_k W_2^k$. □

<u>5.1.4.</u> We end this section by giving further examples of PDOs.

<u>Example 5.1.16.</u> The singular integro-differential operator

$$A = \sum_{0 \le j \le m} (a_j + b_j S_\Gamma) D^j \ , \ a_j, b_j \in C^\infty$$

is a PDO of order m on Γ with symbol

$$\sum_{0 \le j \le m} (a_j + b_j \xi/|\xi|)(i\xi)^j \ .$$

This is an immediate consequence of Examples 5.1.4 and 5.1.5 and Theorem 5.1.11.

Without loss of generality we now suppose $0 \in \Omega$, Ω being the bounded domain with boundary Γ. We consider the following integral operators with logarithmic kernels:

$$\begin{aligned} (T^+u)(t) &= (2\pi i)^{-1} \int_\Gamma k^+(t,\tau) \ln(1-t/\tau) u(\tau)\, d\tau \ , \\ (T^-u)(t) &= (2\pi i)^{-1} \int_\Gamma k^-(t,\tau) \ln(1-\tau/t) u(\tau)\, d\tau \ , \ t \in \Gamma \ , \end{aligned} \tag{5.1.20}$$

where $k^{\pm} \in C^{\infty}(\Gamma \times \Gamma)$ and $\ln(1-t/\tau)(\ln(1-\tau/t))$ for fixed $t \in \Gamma$ denotes the branch of the logarithm which is continuous for $\tau \neq t$ and $\tau \in \mathbb{C} \setminus \Omega$ $(\tau \in \bar{\Omega})$ and vanishes for $\tau = \infty$ $(\tau = 0)$.

Expanding $k^{\pm}$ into Taylor series at $\tau = t$, for any $N \in \mathbb{N}_0$, we obtain

$$k^{\pm}(t,\tau) = \sum_{0 \leq j \leq N} a_j^{\pm}(t)(\tau - t)^j/j! + a_{N+1}^{\pm}(t,\tau)(\tau - t)^{N+1} \qquad (5.1.21)$$

with certain functions $a_j^{\pm} \in C^{\infty}(\Gamma)$ and $a_{N+1}^{\pm} \in C^{\infty}(\Gamma \times \Gamma)$. Note that $a_o^{\pm}(t) = k^{\pm}(t,t)$.

<u>Example 5.1.17.</u> The operators $T^{\pm}$ defined in (5.1.2o) are PDOs of order -1 on Γ with symbols

$$\sigma^{\pm} = \mp \sum_{j \geq 0} (-1)^j a_j^{\pm}(t) h(\pm \xi)[(i\xi)^{-1} \circ t'(x)]^{j+1} , \qquad (5.1.22)$$

where $[a]^j$ denotes the composition $a \circ \ldots \circ a$ (j factors).

In order to prove these assertions, we need two lemmas.

<u>Lemma 5.1.18.</u> For any $u \in C^{\infty}$, the following relations hold:

$$D_t T^+ u = (2\pi i)^{-1} \int_{\Gamma} (\partial k^+/\partial t) \ln(1-t/\tau) u(\tau) d\tau - 2^{-1} k^+(t,t)(I+S_{\Gamma})u$$
$$+ (2\pi i)^{-1} \int_{\Gamma} [k^+(t,t) - k^+(t,\tau)](\tau - t)^{-1} u(\tau) d\tau , \qquad (5.1.23)$$
$$D_t T^- u = (2\pi i)^{-1} \int_{\Gamma} (\partial k^-/\partial t) \ln(1-\tau/t) u(\tau) d\tau + 2^{-1} k^-(t,t)(I-S_{\Gamma})u$$
$$+ (2\pi i)^{-1} \int_{\Gamma} \{[k^-(t,t) - k^-(t,\tau)](\tau - t)^{-1} - t^{-1} k^-(t,\tau)\} u(\tau) d\tau ,$$

where $D_t = d/dt = (t'(x))^{-1} D$.

<u>Proof.</u> Similarly as in the proof of Lemma 1.5 in Prößdorf [1, Chap. 6] one first differentiates the integrals

$$\int_{\Gamma \setminus \Gamma_{\varepsilon}} k^+(t,\tau) \ln(1-t/\tau) u(\tau) d\tau , \quad \int_{\Gamma \setminus \Gamma_{\varepsilon}} k^-(t,\tau) \ln(1-\tau/t) u(\tau) d\tau ,$$

where $\Gamma_{\varepsilon} \subset \Gamma$ is an arc of length ε and midpoint t, and letting $\varepsilon \to 0$, one then obtains (5.1.23). □

Similarly, using an integration by parts, one can prove

<u>Lemma 5.1.19.</u> For any $u \in C^{\infty}$, we have

$$T^+ D_t u = -(2\pi i)^{-1} \int_{\Gamma} (\partial k^+/\partial \tau) \ln(1-t/\tau) u(\tau) d\tau -$$
$$- 2^{-1} k^+(t,t)(I+S_{\Gamma})u + (2\pi i)^{-1} \int_{\Gamma} [k^+(t,t) - \qquad (5.1.24)$$
$$k^+(t,\tau) t/\tau](\tau - t)^{-1} u(\tau) d\tau ,$$

$$T^-D_t u = -(2\pi i)^{-1}\int_\Gamma (\partial k^-/\partial\tau)\ln(1-\tau/t)u(\tau)d\tau + \quad (5.1.24)$$
$$+ 2^{-1}k^-(t,t)(I-S_\Gamma)u + (2\pi i)^{-1}\int_\Gamma [k^-(t,t)-k^-(t,\tau)](\tau-t)^{-1} u(\tau)d\tau .$$

Now it follows from Lemma 5.1.18 that $T^\pm \in L(W_2^k, W_2^{k+1})$ for all $k \in \mathbb{N}_0$, hence $T^\pm \in L(C^\infty)$. Indeed, since the first terms on the right-hand sides of (5.1.23) take the form (5.1.2o) again and thus are integral operators with weakly singular kernels, we obtain $T^\pm, D_tT^\pm \in L(W_2^0)$, hence $T^\pm \in L(W_2^0, W_2^1)$. Applying formulas (5.1.23) to $D_tT^\pm$ and using the facts that $S_\Gamma \in L(W_2^k)$ for any k and the last terms on the right-hand sides of (5.1.23) are smoothing operators, we deduce $T^\pm \in L(W_2^1, W_2^2)$. By iteration, $T^\pm \in L(W_2^k, W_2^{k+1})$, $k \in \mathbb{N}$ Q.E.D.

Furthermore, by Lemma 5.1.19 and (5.1.21) we have

$$T^\pm D_t \equiv T_1^\pm \mp a_0^\pm P^\pm ,$$

where $T_1^\pm$ denote the first operators on the right-hand sides of (5.1.24). Setting $V = Op((i\xi)^{-1} \circ t'(x))$, one obtains that

$$T^\pm \equiv T^\pm D_t V \equiv T_1^\pm V \mp a_0^\pm P^\pm V . \quad (5.2.15)$$

Indeed, we have $D_tV \equiv I$ by Theorem 5.1.11, and $T^\pm \in L(C^\infty)$ together with Theorem 5.1.8 and Remark 5.1.9 implies that $T^\pm(D_tV-I)$ are smoothing operators.

Applying formulas (5.1.24) again, we obtain an analogous representation of $T_1^\pm$, and by iteration it follows from (5.1.25) and (5.1.21) that

$$T^\pm \equiv \mp \sum_{0\le j<N} (-1)^j a_j^\pm P^\pm V^{j+1} + (-1)^N T_N^\pm V^N, \; N=1,2,\ldots, \quad (5.1.26)$$

where

$$T_N^+u = (2\pi i)^{-1}\int_\Gamma (\partial^N k^+/\partial\tau^N)\ln(1-t/\tau)u(\tau)d\tau ,$$
$$T_N^-u = (2\pi i)^{-1}\int_\Gamma (\partial^N k^-/\partial\tau^N)\ln(1-\tau/t)u(\tau)d\tau .$$

We have seen above that $T_N^\pm \in L(W_2^k, W_2^{k+1})$ for any $k \in \mathbb{N}_0$. Hence

$$T_N^\pm V^N \in L(W_2^k, W_2^{k+N+1}) , \; k=-N,-N+1,\ldots$$

in view of $V^N \in OPCS^{-N}$. Therefore, by (5.1.26) and Theorem 5.1.11 we get $T^\pm - Op(\sigma^\pm) \in OPCS^l$ for any $l \in \mathbb{R}$ with $\sigma^\pm$ given by (5.1.22). This proves the result of Example 5.1.17.

5.2. Gårding-Melin inequalities

Retaining the notation of the preceding section, we now prove some a priori estimates for PDOs on a closed curve which are basic for the investigation of degenerate PDOs in Sec. 5.3. Let U be an open arc on Γ, $A \in OPCS^1$ with symbol $a_1 + a_{1-1} + \dots$ and $a_1^{\pm}(t) = a_1(t, \pm 1)$. The subprincipal symbol of A is defined by

$$a'_{1-1} = a_{1-1} - (2i)^{-1} D_x D_\xi a_1 .$$

Theorem 5.2.1. If $a_1^+(t) \neq 0$ for all $t \in \bar{U}$, then for any $k \in \mathbb{R}$ there exists a constant $c = c(k) > 0$ such that

$$\| Au \|_k + \| u \|_{k+1-1} \geq c \| P^+ u \|_{k+1}, \quad u \in C_0^\infty(U). \qquad (5.2.1)$$

If $a_1^-(t) \neq 0$ $(t \in \bar{U})$, then (5.2.1) holds with P^+u replaced by P^-u.

Proof. By Theorem 5.1.7 we have

$$A = (a_1^+ P^+ + a_1^- P^-)\Lambda^1 + A_1, \quad A_1 \in OPCS^{1-1} .$$

If $a_1^+(t) \neq 0$ $(t \in \bar{U})$, then there is some open arc $V \supset \bar{U}$ such that $a_1^+(t) \neq 0$ $(t \in V)$. Choose $\chi \in C^\infty(\Gamma)$ such that $\chi = 1$ in U and $\chi = 0$ in some neighborhood of $\Gamma \setminus V$. Setting $b = \chi / a_1^+$ on V and $b = 0$ on $\Gamma \setminus V$ and $B = bP^+\Lambda^{-1}$, by Theorem 5.1.11 one obtains

$$BA = P^+ b a_1^+ I + T, \quad T \in OPCS^{-1}, \quad b a_1^+ = 1 \text{ on } U.$$

Together with Theorem 5.1.8, this implies

$$\| P^+u \|_{k+1} \leq \| BAu \|_{k+1} + \| Tu \|_{k+1} \leq c(\| Au \|_k + \| u \|_{k+1-1}), u \in C_0^\infty(U) ,$$

hence (5.2.1). The proof of the second assertion is analogous. □

Theorem 5.2.2. If $\mathrm{Re}\, a_1^+(t) > 0$ $(t \in \bar{U})$, then for any $k \geq 1$ there exists a constant $c = c(k) > 0$ such that

$$\mathrm{Re}(Au, P^+u)_0 + \| u \|^2_{(1-k)/2} \geq c \| P^+u \|^2_{1/2}, \quad u \in C_0^\infty(U) . \qquad (5.2.2)$$

If $\mathrm{Re}\, a_1^-(t) > 0$ $(t \in \bar{U})$, then (5.2.2) holds with P^+u replaced by P^-u.

Proof. Let $\mathrm{Re}\, a_1^+(t) > 0$ $(t \in \bar{U})$, for example. It is sufficient to show (5.2.2) for k=1, since for arbitrary $\varepsilon > 0$ and $k > 1$ there exists $c(\varepsilon, k) > 0$ such that

$$\| u \|^2_{(1-1)/2} \leq \varepsilon \| u \|^2_{1/2} + c(\varepsilon, k) \| u \|^2_{(1-k)/2}, \quad u \in C^\infty.$$

This is an easy consequence of the definition (5.1.2) of the norm in W_2^k. Furthermore, note that for $T \in OPCS^{l-1}$

$$\operatorname{Re}(Tu,u)_0 \geq - c \, \|u\|^2_{(l-1)/2} \, , \quad u \in C^\infty$$

with some $c > 0$ since

$$|(Tu,u)_0| \leq c_1 \|Tu\|_{(l-1)/2} \|u\|_{(l-1)/2} \leq c_2 \|u\|^2_{(l-1)/2}$$

(in the following let $c, c_1, \ldots$ denote positive constants independent of u). In view of the representation

$$A = \Lambda^{1/2}(a_l^+ P^+ + a_l^- P^-)\Lambda^{1/2} + T, \quad T \in OPCS^{l-1} \, ,$$

it is thus sufficient to prove (5.2.2) for the operator $B = a_l^+ P^+ + a_l^- P^-$ which belongs to $OPCS^0$. Moreover, since

$$\operatorname{Re}(Bu,u)_0 = \operatorname{Re}(2^{-1}(B+B^*)u,u)_0$$

and $2^{-1}(B+B^*)$ has the principal symbol $\operatorname{Re} a_l^+ h(\xi) + \operatorname{Re} a_l^- h(-\xi)$ (see (5.1.14) and (5.1.17)), we can assume that $a_l^+(t) > 0$ $(t \in \bar{U})$. Choose an open arc $V \supset \bar{U}$ such that $a_l^+(t) > 0$ $(t \in V)$, a function χ as in the proof of Theorem 5.2.1 and $a \in C^\infty$ such that $a > 0$ on Γ and $a = a_l^+$ on $\operatorname{supp} \chi$. Using Theorems 5.1.11 and 5.1.12, we then obtain

$$(B\chi u, P^+\chi u)_0 = (aP^+\chi u, P^+\chi u)_0 + (K\chi u, \chi u)_0 \, , \quad u \in C^\infty,$$

where K is a smoothing operator. This implies the inequality

$$\operatorname{Re}(Bu,P^+u)_0 + \|u\|^2_{-1/2} \geq c \, \|P^+u\|^2_0, \quad u \in C_0^\infty(U) \qquad \text{Q.E.D.} \quad \square$$

A is called strongly elliptic if its principal symbol satisfies $\operatorname{Re} a_l(t,\xi) > 0$ for all $t \in \Gamma$ and $\xi = \pm 1$.

Corollary 5.2.3. (Gårding's inequality) If A is strongly elliptic, then

$$\operatorname{Re}(Au,u)_0 + \|u\|^2_{(l-1)/2} \geq c \, \|u\|^2_{l/2} \, , \quad u \in C^\infty. \tag{5.2.3}$$

Proof. As in Theorem 5.2.2 we get the inequalities

$$\operatorname{Re}(Au,P^\pm u)_0 + \|u\|^2_{(l-1)/2} \geq c \, \|P^\pm u\|^2_{l/2}, \quad u \in C^\infty$$

which immediately imply (5.2.3) in view of (5.1.2) and (5.1.14). $\square$

Theorem 5.2.4. If $\operatorname{Re} a_l^+ \geq 0$ on Γ and $\operatorname{Re} a^{+}_{l-1}(t,1) > 0$ when $\operatorname{Re} a_l^+(t) = 0$ and $t \in \bar{U}$, then

$$\mathrm{Re}(Au,P^+u)_0 + \|u\|^2_{(l-2)/2} \geq c\,\|P^+u\|^2_{(l-1)/2},\ u \in C_0^\infty(U)\ . \tag{5.2.4}$$

If $\mathrm{Re}\ a_l^- \geq 0$ on Γ and $\mathrm{Re}\ a'_{l-1}(t,-1) > 0$ when $\mathrm{Re}\ a_l^-(t) = 0$ and $t \in \bar{U}$, then (5.2.4) is valid with P^+u replaced by P^-u.

<u>Proof.</u> 1. We shall show inequality (5.2.4). First we reduce the assertion to the case when a_l and a_{l-1} are real. By Theorem 5.1.12, $(A+A^*)/2$ has the symbol

$$\begin{aligned}&\mathrm{Re}\ a_l + (\mathrm{Re}\ a_{l-1}+(2i)^{-1}D_xD_\xi\, a_l)+\ldots = \mathrm{Re}\ a_l \\ &+ (\mathrm{Re}\ a'_{l-1}+i\ \mathrm{Im}\{(2i)^{-1}D_xD_\xi\, a_l\})+\ldots\ .\end{aligned} \tag{5.2.5}$$

Furthermore, if $C \in OPCS^{l-1}$ has a purely imaginary principal symbol, then $(C+C^*)/2 \in OPCS^{l-2}$. Consequently, (5.2.5) implies

$$\mathrm{Re}(Au,u)_0 = \mathrm{Re}\{(A_1u,u)_0 + (Tu,u)_0\}$$

with

$$A_1 = \mathrm{Op}(\mathrm{Re}\ a_l+\mathrm{Re}\ a'_{l-1}),\ T \in OPCS^{l-2}\ .$$

Thus it is sufficient to prove (5.2.4) when $a_l^+ \geq 0$ on Γ and $a_{l-1}^+(t) > 0$ when $a_l^+(t) = 0$ and $t \in \bar{U}$.

2. We now reduce the assertion to the case $l=1$. Consider the operator $\Lambda^{(1-l)/2}A\Lambda^{(1-l)/2}$. Its principal symbol is $|\xi|^{1-l}a_l$ and the next lower term is the sum of $|\xi|^{1-l}a_{l-1}$ and a purely imaginary term. Therefore it suffices to verify (5.2.4) for an operator

$$A = (a_1^+\Lambda + a_0^+)P^+ + (a_1^-\Lambda + a_0^-)P^-$$

satisfying $a_1^+ \geq 0$ on Γ and $a_0^+(t) > 0$ when $a_1^+(t) = 0$ and $t \in \bar{U}$.

3. Choose a finite covering $\{U_j : j=1,\ldots,m\}$ of $\bar{U}$ by open arcs on Γ such that $a_1^+ > 0$ or $a_0^+ > 0$ on $\bar{U}_j$ $(j=1,\ldots,m)$. In the first case, by Theorem 5.2.2 (for $k=2$) we obtain

$$\mathrm{Re}(Au,P^+u)_0 + \|u\|^2_{-1/2} \geq c\|P^+u\|^2_{1/2} \geq c\|P^+u\|^2_0,\ u \in C_0^\infty(U_j). \tag{5.2.6}$$

In the second case, we use the equality

$$\begin{aligned}\mathrm{Re}(Au,P^+u)_0 &= \mathrm{Re}\{(a_0^+P^+u,P^+u)_0 + (a_1^+\Lambda^{1/2}P^+u,\Lambda^{1/2}P^+u)_0 \\ &+ ([a_1^+I,\Lambda^{1/2}]\Lambda^{1/2}P^+u,P^+u)_0\}\ .\end{aligned} \tag{5.2.7}$$

For any $u \in C_0^\infty(U_j)$, the first term on the right-hand side of (5.2.7) can be estimated from below by $c\|P^+u\|_0^2 - \|u\|_{-1/2}^2$ (cf. Theorem 5.2.2), the second term is non-negative since $a_1^+ \geq 0$ on Γ and the last term can be estimated from below by $-c\|u\|_{-1/2}^2$ since $[a_1^+ I, \Lambda^{1/2}]\Lambda^{1/2} \in OPCS^0$ and the principal symbol of this operator is purely imaginary. Therefore one obtains (5.2.6) for all j.

4. We further choose functions $\chi_j \in C^\infty(\Gamma)$ $(j=1,\dots,m)$ satisfying $\operatorname{supp}\chi_j \subset U_j$ and $\sum \chi_j^2 \equiv 1$ on $\bar{U}$. Estimates (5.2.5) for $j=1,\dots,m$ and the calculus of PDOs then imply

$$\operatorname{Re}(Au,P^+u)_0 = \operatorname{Re}(Au,P^+\sum\chi_j^2 u)_0 = \sum \operatorname{Re}(AP^+\chi_j u, P^+\chi_j u)_0$$
$$- c\|u\|_{-1/2}^2 \geq c_1\sum\|P^+\chi_j u\|_0^2 - c_2\|u\|_{-1/2}^2$$
$$\geq c_3\|P^+u\|_0^2 - c_4\|u\|_{-1/2}^2, \quad u \in C_0^\infty(U) \qquad \text{Q.E.D.} \ \square$$

Corollary 5.2.5. (Melin's inequality) If

$$\operatorname{Re} a_1 \geq 0 \quad \text{on} \quad \Gamma \times \{\pm 1\},$$
$$\operatorname{Re} a_{l-1}^{!} > 0 \quad \text{on} \quad \{(t,\pm 1) : t \in \Gamma, \ \operatorname{Re} a_1^{\pm}(t) = 0\},$$

then

$$\operatorname{Re}(Au,u)_0 + \|u\|_{(l-2)/2}^2 \geq c\|u\|_{(l-1)/2}^2, \quad u \in C^\infty. \tag{5.2.8}$$

Proof. As in Theorem 5.2.4 one can deduce the estimates

$$\operatorname{Re}(Au,P^{\pm}u)_0 + \|u\|_{(l-2)/2}^2 \geq c\|P^{\pm}u\|_{(l-1)/2}^2, \quad u \in C^\infty$$

which imply (5.2.8). $\square$

5.3. The index for a class of degenerate pseudodifferential operators

5.3.1. Consider a PDO $A \in OPCS^l$ with symbol $a_0 + a_{l-1} + \dots$ and let $a_{l-j}^{\pm}(t) = a_{l-j}(t,\pm 1)$, $j=0,1,\dots$. We have seen in Theorem 5.1.14 that $A \in L(W_2^{k+l}, W_2^k)$ is not a Fredholm operator if A is not elliptic. In this section we shall study Fredholm property and index of the closed operators

$$A : W_2^{k+l-1} \to W_2^k, \quad D_k(A) = \{u \in W_2^{k+l-1} : Au \in W_2^k\}. \tag{5.3.1}$$

The domain of definition $D_k(A)$ endowed with the graph norm

$$\|u\|_{k+l-1} + \|Au\|_k$$

is a Banach space which we denote by $W_{2,A}^k$, and obviously $A \in L(W_{2,A}^k, W_2^k)$.

Lemma 5.3.1. For any $k \in \mathbb{R}$, C^∞ is dense in $W^k_{2,A}$.

Proof. Let first $A \in OPCS^1$ and $k=0$. Using a Friedrichs' mollifier, for any $u \in W^0_{2,A}$ one can construct a sequence $\{u_n\} \subset C^\infty$ such that $u_n \to u$ and $Au_n \to Au$ $(n \to \infty)$ in L_2 (see Taylor [1, Chap. 2, Prop. 7.4]).

In the general case we pass to the operator $B = \Lambda^k A \Lambda^{1-l-k}$ with Λ^k defined in (5.1.7). For given $u \in W^k_{2,A}$, set $v = \Lambda^{k+l-1}u$. Then $v \in W^0_{2,B}$ and there exists a sequence $\{v_n\} \subset C^\infty$ such that $v_n \to v, Bv_n \to Bv$ $(n \to \infty)$ in L_2. Setting $u_n = \Lambda^{1-l-k}v_n$, we obtain $u_n \to u$ in W_2^{k+l-1} and $Au_n \to Au$ in W_2^k as $n \to \infty$. □

Lemma 5.3.2. $\|u\|_{W^k_{2,A}} = \|u\|_{k+l-1} + \|a_l^+ DP^+u\|_{k+l-1} + \|a_l^- DP^- u\|_{k+l-1}$

is an equivalent norm in $W^k_{2,A}$.

Proof. Since $\|u\|_k^2 = \|P^+u\|_k^2 + \|P^-u\|_k^2$, the graph norm in $D_k(A)$ is equivalent to

$$\|u\|_{k+l-1} + \|P^+Au\|_k + \|P^-Au\|_k .$$

Using the representations

$$P^\pm A = \Lambda^{l-1}(\pm i a_l^\pm) DP^\pm + A^\pm , \; A^\pm \in OPCS^{l-1}$$

(cf. (5.1.15) and Theorem 5.1.11), we get the conclusion. □

We now give a sufficient condition for (5.3.1) to be a Fredholm operator and an index formula. Define the functions

$$a^k = a_l + a^k_{l-1} , \; a^k_{l-1} = a_{l-1} + (k-1/2)(i\xi)^{-1} D_x a_l . \tag{5.3.2}$$

Theorem 5.3.3. Suppose there exist functions $\varrho^\pm \in C^\infty$ such that $\varrho^\pm(t) \neq 0$ $(t \in \Gamma)$ and

$$\operatorname{Re} \varrho^\pm a_l^\pm \geq 0 \text{ on } \Gamma , \; \operatorname{Re} \varrho^\pm a^k_{l-1}(t,\pm 1) > 0 \text{ on } \Sigma^\pm , \tag{5.3.3}$$

where $\Sigma^\pm = \{t \in \Gamma : a_l^\pm(t) = 0\}$. Then $A \in L(W^k_{2,A}, W^k_2)$, or equivalently (5.3.1), is a Fredholm operator with index

$$\operatorname{ind} A = (2\pi)^{-1} \{\arg a^k(t,-\xi)/a^k(t,\xi)\}_\Gamma \tag{5.3.4}$$

for sufficiently large ξ.

We first reduce the assertion to the case $\varrho^\pm = 1$ and $l-1=k=0$. Using the operator $C = (\varrho^+)^{-1}P^+ + (\varrho^-)^{-1}P^-$, Theorem 5.1.11 and (5.1.10), we obtain the equality

$$A = CB + T \tag{5.3.5}$$

where T is a smoothing operator and B a PDO with symbol $b_l + b_{l-1} + \dots$,

$$b_l = [\varrho^+ a_l^+ h(\xi) + \varrho^- a_l^- h(-\xi)]\, |\xi|^l$$

$$b_{l-1} = [\varrho^+ a_{l-1}^+ h(\xi) + \varrho^- a_{l-1}^- h(-\xi)]\, |\xi|^{l-1}.$$

By Theorem 5.1.14, $C \in L(W_2^k)$ is a Fredholm operator with index $(2\pi)^{-1} \times$ $\times \{\arg \varrho^+/\varrho^-\}_\Gamma$. Furthermore,

$$\{\arg \varrho^{\pm}(t) a^k(t, \pm\xi)\}_\Gamma = 0 \text{ when } \xi \text{ is sufficiently large.} \tag{5.3.6}$$

Indeed, by (5.3.3) there exist neighborhoods $U^{\pm}$ of $\Sigma^{\pm}$ and $c > 0$ such that

$$\operatorname{Re} \varrho^{\pm} a_{l-1}^k (t, \pm\xi) \geq c\, |\xi|^{l-1} \quad (t \in U^{\pm},\ \xi > 0),$$

$$|\varrho^{\pm} a_l (t, \pm\xi)| \geq c\, |\xi|^l \quad (t \in \Gamma \setminus U^{\pm},\ \xi > 0).$$

Together with $\operatorname{Re} \varrho^{\pm} a_l^{\pm} \geq 0$ on Γ, this implies (5.3.6). Moreover, $W_{2,B}^k =$ $= W_{2,A}^k$ (see Lemma 5.3.2) and the assumptionsof the theorem imply

$$\operatorname{Re} b_l^{\pm} \geq 0 \text{ on } \Gamma,\ \operatorname{Re} b_{l-1}^k(t, \pm 1) > 0 \text{ when } b_l^{\pm}(t) = 0\,, \tag{5.3.7}$$

where $b_l^{\pm} = b_l(t, \pm 1)$ and b_{l-1}^k is defined as in (5.3.2). By virtue of (5.3.5) and (5.3.6), it is thus sufficient to show that $B \in L(W_{2,B}^k, W_2^k)$ is a Fredholm operator with index 0 if (5.3.7) holds.

Consider the operator $E = \Lambda^k B \Lambda^{1-k-l}$ and the commutative diagram

$$\begin{array}{ccc} B : W_{2,B}^k & \longrightarrow & W_2^k \\ \downarrow \Lambda^{k+l-1} & & \downarrow \Lambda^k \\ E : W_{2,E}^o & \longrightarrow & W_2^o \end{array} \tag{5.3.8}$$

where $W_{2,E}^o = W_{2,B}^{1-l} = \{ u \in L_2 : b_l^{\pm} DP^{\pm} u \in L_2 \}$. Since Λ^k is an isomorphic map in (5.3.8), Λ^{k+l-1} is an isomorphism, too. Indeed, Λ^{k+l-1} : $W_2^{k+l-1} \to L_2$ is an isomorphism, and $u \in W_2^{k+l-1}$ and $\Lambda^{k+l-1} u \in W_{2,E}^o$ imply $u \in W_{2,B}^k$. Therefore B is a Fredholm operator if and only if E is Fredholm, and their indices coincide.

Using Theorem 5.1.11, we observe that E has the symbol $e_1 + e_o + \dots$, where

$$e_1 = |\xi|^{1-l} b_l\,,\quad e_o = |\xi|^{1-l} b_{l-1} + k\, |\xi|^{-l} i^{-1} \operatorname{sgn}\xi\ D_x b_l\,.$$

Moreover, $(2i\xi)^{-1} l D_x b_l = (2i)^{-1} D_x D_\xi b_l$. Therefore the subprincipal symbol of E can be written

$$e_o' = |\xi|^{1-l} b_{l-1} + k|\xi|^{-l} i^{-1} \operatorname{sgn}\xi \, D_x b_l - |\xi|^{1-l}(2i)^{-1} D_x D_\xi b_l$$
$$- |\xi|^{-l}(2i)^{-1}(1-l)\operatorname{sgn}\xi \, D_x b_l = |\xi|^{1-l}\{b_{l-1} + (k-1/2)(i\xi)^{-1} D_x b_l\},$$

and we obtain

$$b^k = b_l + b_{l-1}^k = |\xi|^{1-l}(e_1 + e_o') .$$

Thus Theorem 5.3.3 is a consequence of the following

Theorem 5.3.4. Let $A \in OPCS^1$ with principal symbol a_1 and subprincipal symbol a_o', and set $\Sigma^{\pm} = \{t \in \Gamma : a_1(t, \pm 1) = 0\}$. Under the hypotheses

$$\operatorname{Re} a_1(t, \pm 1) \geq 0 \text{ on } \Gamma, \ \operatorname{Re} a_o'(t, \pm 1) > 0 \text{ on } \Sigma^{\pm}, \tag{5.3.9}$$

$A \in L(W_{2,A}^o, L_2)$ is a Fredholm operator with index 0.

Before proving Theorem 5.3.4, we establish the following

Lemma 5.3.5. If (5.3.9) is satisfied, then

$$\|Au\|_o^2 + \|u\|_{-1/2}^2 \geq c \|u\|_o^2, \ u \in C^\infty . \tag{5.3.1o}$$

Proof. We shall verify the inequality

$$\|Au\|_o^2 + \|u\|_{-1/2}^2 \geq c \|P^+u\|_o^2 , \ u \in C^\infty . \tag{5.3.11}$$

Choose a finite system of open arcs U_j which cover Γ such that $a_1(t,1) \neq 0$ or $\operatorname{Re} a_o'(t,1) > 0$ on $\overline{U}_j$ for all j. By Theorems 5.2.1 and 5.2.4, one obtains the estimate

$$\|Au\|_o^2 + \|u\|_{-1/2}^2 \geq c \|P^+u\|_o^2, \ u \in C_o^\infty(U_j) \tag{5.3.12}$$

for any j. Choosing a partition of unity $\{\chi_j^2\}$ as in the fourth step of the proof of Theorem 5.2.4, from (5.3.12) and the PDO calculus we deduce

$$\|Au\|_o^2 + \|u\|_{-1/2}^2 \geq c \textstyle\sum_j \|A\chi_j u\|_o^2 \geq c_1 \sum_j \|P^+\chi_j u\|_o^2$$
$$- c_2 \|u\|_{-1/2}^2 \geq c_3 \|P^+u\|_o^2 - c_4 \|u\|_{-1/2}^2, \ u \in C^\infty$$

which proves (5.3.11). Analogously, (5.3.11) holds with P^+u replaced by P^-u which completes the proof of (5.3.1o). □

Proof of Theorem 5.3.4. We shall show that the closed operators

$$A_\lambda = A+\lambda I : L_2 \to L_2,\ D(A_\lambda) = \{u \in L_2 : Au \in L_2\} \tag{5.3.13}$$

are Fredholm operators with index 0 for all $\lambda \geq 0$. (5.3.9) and Lemma 5.3.5 imply

$$\|A_\lambda u\|_0^2 + \|u\|_{-1/2}^2 \geq c(\lambda)\|u\|_0^2,\ u \in C^\infty,\ \lambda \geq 0. \tag{5.3.14}$$

Furthermore, consider the formal adjoint of A_λ

$$A_\lambda^* = A^* + \lambda I : L_2 \to L_2,\ D(A_\lambda^*) = \{u \in L_2 : A^* u \in L_2\}. \tag{5.3.15}$$

Obviously, $D(A_\lambda) = D(A_\lambda^*) = W_{2,A}^0$ in view of Lemma 5.3.2. By Theorem 5.1.12, A^* has the principal symbol $\bar{a}_1$ and the subprincipal symbol $\bar{a}_0 + (2i)^{-1} D_x D_\xi \bar{a}_1$. Therefore A_λ^* ($\lambda \geq 0$) satisfies the assumptions of Lemma 5.3.5, and we have

$$\|A_\lambda^* u\|_0^2 + \|u\|_{-1/2}^2 \geq c_1(\lambda)\|u\|_0^2,\ u \in C^\infty,\ \lambda \geq 0. \tag{5.3.16}$$

By Lemma 5.3.1, estimates (5.3.14) and (5.3.16) are valid for all $u \in W_{2,A}^0$ and (5.3.15) is the adjoint of the operator (5.3.13) in L_2. Since the embedding $L_2 \subset W_2^{-1/2}$ is compact, it follows from Theorem VII.2.1 in Goldberg [1] that A_λ and A_λ^* ($\lambda \geq 0$) are Φ_+-operators in L_2, hence (5.3.13) is a Fredholm operator for any $\lambda \geq 0$.

Further, there exists $\lambda_0 > 0$ such that $\operatorname{Re}\{a_0'(t,\pm 1) + \lambda_0\} > 0$ on Γ. By Corollary 5.2.5 applied to A_λ and A_λ^*, we then obtain

$$\|A_\lambda u\|_0^2 \geq c\|u\|_0^2,\quad \|A_\lambda^* u\|_0^2 \geq c\|u\|_0^2,\ u \in W_{2,A}^0$$

for all sufficiently large λ. Thus the operators (5.3.13) are invertible for those λ, and A_λ has index 0 for all $\lambda \geq 0$ since the index is a homotopy invariant. □

<u>5.3.2.</u> Under a certain restriction on the principal symbol, we now give a necessary and sufficient condition for the Fredholm property of the operator (5.3.1). We say that a function $b \in C^\infty(\Gamma)$ satisfies condition (R) if for every $t \in \Gamma$ there exist an (open) neighborhood $U \subset \Gamma$ of t and functions $\beta, \tilde{b} \in C^\infty(U)$ such that $b = \beta\tilde{b}$, $|\tilde{b}| = 1$ and $\operatorname{Im}\beta = 0$ in U.

<u>Theorem 5.3.6.</u> Suppose $a_1^\pm$ satisfy condition (R). Then $A \in L(W_{2,A}^k, W_2^k)$ is a Fredholm operator if and only if there exists $\xi_0 > 0$ such that

$$|a^k(t,\xi)|\,|\xi|^{1-1} \geq c > 0,\ t \in \Gamma,\ |\xi| \geq \xi_0. \tag{5.3.17}$$

If (5.3.17) is valid, then (5.3.1) has the index (5.3.4) for $\xi \geq \xi_o$.

Before proving Theorem 5.3.6, we give some remarks on conditions (R) and (5.3.17) and the index formula.

Remark 5.3.7. It is clear that (R) is only a restriction at the points of $\Sigma^{\pm}$ where $a_1^{\pm}$ vanish to infinite order.

Remark 5.3.8. If $a_1^{\pm}$ satisfy condition (R), we have $a_1^{\pm} = \alpha^{\pm} a^{\pm}, |a^{\pm}| = 1$ and $\operatorname{Im} \alpha^{\pm} = 0$ with certain smooth functions $\alpha^{\pm}, a^{\pm}$ in some neighborhood of any point $t_o \in \Gamma$. Then (5.3.17) holds if and only if, for any $t_o \in \Sigma^{\pm}$, one of the following conditions is satisfied:

(i) $a_1^{\pm} \equiv 0$ in some neighborhood of t_o and $a_{1-1}^{\pm}(t_o) \neq 0$.

(ii) $a_1^{\pm} \not\equiv 0$ in every neighborhood of t_o and $\operatorname{Im} \mu^{\pm} \neq 0$, where $\mu^{\pm} = a_{1-1}^{k}(t_o, \pm 1)/a^{\pm}(t_o)$.

(iii) $a_1^{\pm} \not\equiv 0$ and $\alpha^{\pm} \geq 0$ (resp. ≤ 0) in every sufficiently small neighborhood of t_o and $\operatorname{Im} \mu^{\pm} = 0$, $\operatorname{Re} \mu^{\pm} > 0$ (resp. < 0).

Proof. If for any $t_o \in \Sigma^{\pm}$ one of conditions (i) to (iii) is satisfied, then there exist neighborhoods $U_o^{\pm}$ of t_o such that

$$\varliminf_{\xi \to \infty} \inf_{t \in U_o^{\pm}} |a^k(t, \pm \xi)| \, |\xi|^{1-1} > 0 ,$$

and covering $\Sigma^{\pm}$ by a finite system of such neighborhoods $U_o^{\pm}$, one can find neighborhoods $U^{\pm}$ of $\Sigma^{\pm}$ such that

$$\varliminf_{\xi \to \infty} \inf_{t \in U^{\pm}} |a^k(t, \pm \xi)| \, |\xi|^{1-1} > 0 .$$

Together with

$$|a_1(t, \pm \xi)| \geq c |\xi|^{1}, \quad t \in \Gamma \setminus U^{\pm}, \; \xi > 0 ,$$

this implies (5.3.17). On the other hand, if none of conditions (i) to (iii) is satisfied, then there are sequences $t_n \to t_o$, $|\xi_n| \to \infty$ such that $a^k(t_n, \xi_n) \to 0$ $(n \to \infty)$ which contradicts (5.3.17). □

Remark 5.3.9. Under the assumptions (R) and (5.3.17), the index of (5.3.1) does not depend on the values of $a_{1-1}^{k}(t, \pm 1)$ on $\Gamma \setminus \Sigma^{\pm}$. Indeed, if $b^{\pm} \in C^{\infty}$ coincide with $a_{1-1}^{k}(t, \pm 1)$ on $\Sigma^{\pm}$, then

$$\{\arg a^k(t, \pm \xi) |\xi|^{1-1}\}_{\Gamma} = \{\arg[a_1(t, \pm \xi) |\xi|^{1-1} + b^{\pm}(t)]\}_{\Gamma}$$

for sufficiently large ξ, since for any $\varepsilon > 0$ there exist neighborhoods $U^{\pm}$ of $\Sigma^{\pm}$ such that $|b^{\pm}-a^{k}_{l-1}(t,\pm 1)| \leq \varepsilon$ in $U^{\pm}$ and $|a^{\pm}_{l}| \geq c > 0$ in $\Gamma \setminus U^{\pm}$.

<u>Corollary 5.3.10.</u> If $a^{\pm}_{l}$ satisfy condition (R) and have only zeros of order ≥ 2 on Γ, then (5.3.17) is equivalent to the condition

$$\varliminf_{\xi \to \infty} \inf_{t \in \Gamma} |a(t,\xi)|\,|\xi|^{1-l} > 0 \ , \ a = a_l + a_{l-1}. \tag{5.3.18}$$

Furthermore, for any $k \in \mathbb{R}$, the index formula (5.3.4) can then be replaced by

$$\operatorname{ind} A = (2\pi)^{-1} \{\arg a(t,-\xi)/a(t,\xi)\}_{\Gamma} \ , \tag{5.3.19}$$

where ξ is sufficiently large.

<u>Proof of Theorem 5.3.6.</u>

1\. Sufficiency of condition (5.3.17). By Theorem 5.3.3, it is sufficient to verify the existence of functions $\varrho^{\pm}$ having the properties stated there. We shall construct ϱ^{+}, for example. First we show that, for every $t_o \in \Gamma$, there exists a neighborhood $U \subset \Gamma$ of t_o and a function $\varrho_U \in C^{\infty}(U)$ such that $\varrho_U(t) \neq 0$, $\operatorname{Re} \varrho_U a^{+}_{l} \geq 0$ and, if $t_o \in \Sigma^{+}$, $\operatorname{Re} \varrho_U a^{k}_{l-1}(t,1) > 0$ $(t \in U)$. This is obvious when $t_o \bar{\in} \Sigma^{+}$. For $t_o \in \Sigma^{+}$, the assertion follows easily from Remark 5.3.8 by choosing $\varrho_U = 1/a^{+}_{l-1}$ in case (i), $\varrho_U = (\operatorname{sgn} \operatorname{Im} \mu^{+})/ia^{+}$ in case (ii) and $\varrho_U = 1/a^{+}$ in case (iii).

We now cover Γ by a finite system $\{U_j\}$ of these neighborhoods such that any triplet of them has an empty intersection, and denote the corresponding function ϱ_U by ϱ_j. We may further assume that $\operatorname{Re} \varrho_j a^{k}_{l-1}(t,1) > 0$ in U_j if a^{+}_{l} vanishes at some point in U_j. Choosing a partition of unity $\{\chi_j\}$ such that $\operatorname{supp} \chi_j \subset U_j$, it is easy to check that $\varrho^{+} = \sum_j \chi_j \varrho_j$ has the desired properties. Indeed, if $t \in U_1 \cap U_2 \neq \emptyset$, for example, then

$$\varrho^{+} = \chi_1 \varrho_1 + (1-\chi_1)\varrho_2 \neq 0 \ , \ \operatorname{Re} \varrho^{+} a^{+}_{l} \geq 0$$

and, if $\Sigma^{+} \cap U_1 \cap U_2 \neq \emptyset$, $\operatorname{Re} \varrho^{+} a^{k}_{l-1}(t,1) > 0$ $(t \in U_1 \cap U_2)$. □

2\. Necessity of condition (5.3.17). As in the proof of Theorem 5.3.3 we may assume that $A \in OPCS^{1}$ and $k=0$. Suppose (5.3.17) is violated and the closed operator $A : L_2 \to L_2$, $D(A) = \overset{\circ}{W}_{2,A}$ is Fredholm. Then there exist sequences $\{t_n\} \subset \Gamma$ and $\{\xi_n\} \subset \mathbb{R}$ such that $|\xi_n| \to \infty$ and $|a_1(t_n, \xi_n) + a_0'(t_n, \xi_n)| \to 0$ as $n \to \infty$. We may of course assume that $t_n \to t_0$ for some $t_0 \in \Gamma$. Let further $\xi_n \to \infty$; the case $\xi_n \to -\infty$ is analogous. By condition (R), we have $a_1^+ = \alpha^+ a^+$, $|a^+| = 1$ and $\operatorname{Im} \alpha^+ = 0$ with smooth functions α^+, a^+ in some neighborhood of t_0. Then, by virtue of Remark 5.3.8, the following three cases may occur:

(i) $\alpha^+ \equiv 0$ in some neighborhood of t_0 and $a_0^+(t_0) = 0$.

(ii) α^+ takes positive as well as negative values in every neighborhood of t_0, $\alpha^+(t_0) = 0$ and $\operatorname{Im} \mu = 0$, $\mu = a_0'(t_0, 1)/a^+(t_0)$.

(iii) $\alpha^+ \geq 0$, $\alpha^+ \not\equiv 0$ in every sufficiently small neighborhood of t_0, $\alpha^+(t_0) = 0$ and $\gamma \in (-\infty, 0]$, $\gamma = a_0^+(t_0)/a^+(t_0)$.

Since the embedding $L_2 \subset W_2^{-1}$ is compact and A is a Φ_+-operator, we have the estimate (cf. Goldberg [1, Th. VII.2.1])

$$\| Au \|_0 + \| u \|_{-1} \geq c \| u \|_0 \, , \; u \in C^\infty. \tag{5.3.20}$$

Set $A_1 = (\alpha^+ \Lambda + \gamma) P^+ + P^-$. Since $AP^+ = (a_1^+ \Lambda + a_0^+) P^+ + T$ with some $T \in OPCS^{-1}$ and $A_1 P^- = P^-$, (5.3.20) implies

$$\| A_1 u \|_0 + \| u \|_{-1} \geq c_1 \| u \|_0 \, , \; u \in C_0^\infty(U) \, , \tag{5.3.21}$$

where U is a small neighborhood of t_0. In case (i) it is now easily seen that (5.3.21) cannot hold since $\gamma = 0$ and $\alpha^+ \equiv 0$ in U.

In case (ii) we consider the operator

$$B = (\alpha^+ \Lambda + b) P^+ + P^-, \quad b = \gamma - (2i)^{-1} \{ (D\alpha^+)(t_0) - (D\alpha^+)(t) \} .$$

Then (5.3.21) implies

$$\| Bu \|_0 + \| u \|_{-1} \geq c_2 \| u \|_0 \, , \; u \in C_0^\infty(U) \tag{5.3.22}$$

when U is small enough. We may choose three points $t_1 \prec t^* \prec t_2$ in U such that $\alpha^+(t_1) < 0$, $\alpha^+(t^*) = 0$ and $\alpha^+(t_2) > 0$, for example. Extend $\alpha^+ (t \in \widehat{t_1 t_2})$ to a function $\alpha \in C^\infty(\Gamma)$ such that $\alpha \; (t \in \Gamma)$ describes a simple closed curve in the half-plane $\operatorname{Im} t \geq 0$ and $\operatorname{Im} \alpha > 0$ $(t \notin \widehat{t_1 t_2})$. We shall show that $C = (\alpha \Lambda + b) P^+ + P^-$ is not a Φ_+-operator in L_2 so that estimate (5.3.20) cannot hold with A replaced by C. Therefore, using Theorem 5.2.1 and the facts that α does not vanish in some neighborhood of $\widehat{t_2t_1}$ and $\alpha = \alpha^+$ on $\widehat{t_1t_2}$, we observe that (5.3.22) cannot be valid, hence a contradiction.

Consider the operators $C_\varepsilon = C + i\varepsilon P^+$, $\varepsilon \in \mathbb{R}$. C_ε has the principal symbol

$$[b-(2i)^{-1}D\alpha + i\varepsilon]\, h(\xi) + h(-\xi)\ .$$

For $t \in \widehat{t_1t_2}$, $b-(2i)^{-1}D\alpha = \mu$. Consequently, for $\varepsilon \neq 0$, the operators C_ε satisfy (5.3.17) (with k=0), and for all sufficiently large ξ,

$$(2\pi)^{-1}\{\arg[\alpha\xi + b-(2i)^{-1}D\alpha + i\varepsilon]\}_\Gamma = \begin{cases} 0 & \text{when } \varepsilon > 0, \\ 1 & \text{when } \varepsilon < 0\ . \end{cases}$$

By formula (5.3.4), C_ε has index 0 in L_2 when $\varepsilon > 0$ and index -1 when $\varepsilon < 0$. By virtue of the stability of the index with respect to perturbations with small norm, $C=C_0$ is not a Φ_+-operator in L_2.

In case (iii), let first $\gamma < 0$. To show that estimate (5.3.21) does not hold, it suffices to prove that $B^+ = (\alpha\Lambda+\gamma)P^+ + P^-$ is not a Φ_+-operator in L_2, where $\alpha \in C^\infty(\Gamma)$ is an extension of α^+ $(t \in U)$ such that $\alpha > 0$ in $\Gamma \setminus U$. Arguing by contradiction, we assume that B^+ is a Φ_+-operator in L_2. Since $B^- = P^+ + (-\alpha\Lambda+\gamma)P^-$ is a Fredholm operator in L_2 by the sufficiency part of Theorem 5.3.6, we obtain the estimates

$$\| B^\pm P^\pm u \|_0 + \| u \|_{-1} \geq c \| P^\pm u \|_0\ , \quad u \in C^\infty\ ,$$

which imply

$$\| Bu \|_0 + \| u \|_{-1} \geq c_1 \| u \|_0\ , \quad u \in C^\infty\ , \tag{5.3.23}$$

where

$$B = \alpha D + i\gamma \equiv i\{(\alpha\Lambda+\gamma)P^+ + (-\alpha\Lambda+\gamma)P^-\}\ .$$

We may choose points $t_1 = t(x_1)$ and $t_2 = t(x_2)$ $(-\pi \leq x_1 < x_2 < \pi)$ near t_0 such that $\alpha(x) > 0$ $(x_1 < x \leq x_2)$ and $x = x_1$ is a zero of α of order ≥ 2. Consider the differential operators

$$\begin{aligned} B_\varepsilon &= B+\varepsilon \ : L_2(x_1,x_2) \to L_2(x_1,x_2)\ , \\ D(B_\varepsilon) &= \{ u \in L_2(x_1,x_2) : \alpha Du \in L_2(x_1,x_2), u(x_2) = 0 \}\ . \end{aligned} \tag{5.3.24}$$

Then (5.3.24) has index -1 when $\varepsilon > 0$ and index 0 when $\varepsilon < 0$. The proof of these facts which is analogous to that of Theorem 2.3.2 (with p=2) is left to the reader. Therefore, as in the proof of Theorem

1.3.1 (ii), we see that $B=B_o$ is not normally solvable. On the other hand, if $B_o u_n \to f$ in $L_2(x_1,x_2)$ and $u_n \in D(B_o)$ $(n\to\infty)$, then $\tilde{u}_n \in W^o_{2,B}$ and $B\tilde{u}_n \to \tilde{f}$ in $L_2(\Gamma)$, where the tilde denotes the extension by 0 on $\Gamma\setminus\widehat{t_1t_2}$. By virtue of (5.3.23) and $\tilde{u}_n(x_2) = 0$ for all n, there exists $\tilde{u}\in W^o_{2,B}$ satisfying $B\tilde{u} = \tilde{f}$ and $\tilde{u}(x_2) = 0$ so that $u = \tilde{u}|_{(x_1,x_2)} \in D(B_o)$ and $B_o u = f$, hence a contradiction.

Finally, if $\gamma = 0$ in case (iii) and A is a Fredholm operator in L_2, then $A-\varepsilon P^+$ is Fredholm for small positive ε, hence a contradiction to the assertion just proved. Thus we got a contradiction in each case and the proof of Theorem 5.2.6 is complete. □

5.4. Smoothness of solutions and special cases

5.4.1. We continue to study the degenerate operator (5.3.1) and first give equivalent formulations of condition (5.3.17) and the index formula (5.3.4) when the principal symbol has only isolated zeros of finite order. For $m=1,\dots,n^+$ $(m=1,\dots,n^-)$, let t_m^+ (t_m^-) be distinct points on Γ. Suppose $a_l^\pm$ have the representations

$$a_l^\pm = \beta^\pm \tilde{a}_l^\pm, \quad \beta^\pm = \prod_{1\le m\le n^\pm} (t-t_m^\pm)^{q_m^\pm}, \quad q_m^\pm \in \mathbb{N},$$
$$\tilde{a}_l^\pm(t) \ne 0 \quad (t\in\Gamma). \tag{5.4.1}$$

Let $t_m^\pm = t(x_m^\pm)$, where $x_m^\pm \in [-\pi,\pi)$. We define the numbers

$$\mu_m^\pm(k) = \{a_{l-1}^\pm(x)(x-x_m^\pm)^{q_m^\pm}/a_l^\pm(x) \mp i(k-1/2)D(x-x_m^\pm)^{q_m^\pm}\}|_{x_m^\pm}, \tag{5.4.2}$$

$$\eta_m^\pm(k) = \begin{cases} q_m^\pm/2 & \text{for even } q_m^\pm, \\ (q_m^\pm-1)/2 & \text{for odd } q_m^\pm \text{ and } \operatorname{Im}\mu_m^\pm(k)>0, \\ (q_m^\pm+1)/2 & \text{for odd } q_m^\pm \text{ and } \operatorname{Im}\mu_m^\pm(k)<0. \end{cases} \tag{5.4.3}$$

For $q_m^\pm>1$, $\mu_m^\pm$ and $\eta_m^\pm$ do not depend on k.

Theorem 5.4.1. Assume (5.4.1). Then (5.3.1) is a Fredholm operator if and only if

$$\operatorname{Im}\mu_m^\pm(k) \ne 0 \text{ for odd } q_m^\pm, \ \operatorname{Im}\mu_m^\pm \ne 0 \text{ or } \operatorname{Re}\mu_m^\pm>0$$
$$\text{for even } q_m^\pm, \ m=1,\dots,n^\pm, \tag{5.4.4}$$

and formula (5.3.4) can be written

$$\operatorname{ind} A = (2\pi)^{-1}\{\arg \tilde{a}_1^-/\tilde{a}_1^+\}_\Gamma - \sum_{1\le m\le n^+} \eta_m^+(k) + \\ + \sum_{1\le m\le n^-} \eta_m^-(k) . \tag{5.4.5}$$

To prove this theorem, we need two lemmas.

Lemma 5.4.2. Let $a,b\in C^\infty(\Gamma)$, $\Sigma = \{t\in\Gamma : a(t) = 0\}$, and assume

$$\varliminf_{\xi\to\infty} \inf_{t\in\Gamma} |a(t)\xi + b(t)| > 0 . \tag{5.4.6}$$

Suppose there exist arcs $U_1,U_2\subset\Gamma$ such that $U_1\cap U_2 = \emptyset$, $\Sigma\subset U_1\cup U_2$ and functions $a_j,b_j\in C^\infty(\Gamma)$ $(j=1,2)$ satisfying $a = a_1a_2$ on Γ, $a_1(t)\ne 0$ $(t\in U_2)$, $a_2(t)\ne 0$ $(t\in U_1)$, $b_1 = b/a_2$ in $U_1\cap\Sigma$ and $b_2 = b/a_1$ in $U_2\cap\Sigma$. Then for all sufficiently large ξ,

$$\{\arg(a\xi+b)\}_\Gamma = \{\arg(a_1\xi+b_1)\}_\Gamma + \{\arg(a_2\xi+b_2)\}_\Gamma . \tag{5.4.7}$$

Proof. We have

$$a\xi+b = (a_1\xi+b_1)(a_2\xi+b_2)\xi^{-1} - (b_1b_2\xi^{-1}+a_1b_2+a_2b_1-b) .$$

For any $\varepsilon>0$, there exists some neighborhood U of Σ such that the absolute value of the second term on the right-hand side can be estimated by ε on U when ξ is large enough. Since the absolute value of the first term can be estimated from below by $c\xi$ on $\Gamma\setminus U$ with some $c>0$, we obtain (5.4.7).□

Lemma 5.4.3. Let $t_0 = t(0)\in\Gamma$, $q\in\mathbb{N}$, $a(t) = (t-t_0)^q$ and $b\in\mathbb{C}$. Then (5.4.6) holds if and only if $\operatorname{Im}\mu\ne 0$ for odd q and $\operatorname{Im}\mu\ne 0$ or $\operatorname{Re}\mu>0$ for even q, where $\mu = b(t'(0))^{-q}$. Moreover, under this hypothesis,

$$(2\pi)^{-1}\{\arg(a\xi+b)\}_\Gamma = \begin{cases} q/2 & \text{for even } q , \\ (q-1)/2 & \text{for odd } q \text{ and } \operatorname{Im}\mu>0 , \\ (q+1)/2 & \text{for odd } q \text{ and } \operatorname{Im}\mu<0 \end{cases} \tag{5.4.8}$$

when ξ is sufficiently large.

Proof. The first assertion follows immediately from Remark 5.3.8, since $a(x) = x^q\tilde{a}(x)$ with $\tilde{a}\in C^\infty(U)$ and $\tilde{a}(x)\ne 0$ $(x\in U)$ in some neighborhood U of t_0. It remains to verify formula (5.4.8). We have

$$|t-t_0|^2e^{2i\vartheta} = (t-t_0)^2,\quad \vartheta(t) = \arg(t-t_0),\quad (2\pi)^{-1}\{\arg e^{2i\vartheta}\}_\Gamma = 1.$$

Let q be even. Then

$$a\xi + b = e^{qi\vartheta}a_1,\ a_1 = |t-t_0|^q\xi + b_1,\ b_1 = be^{-qi\vartheta}.$$

Since $b_1(t_0) = |t'(0)|^{q/2}\mu \bar{\in} (-\infty,0]$, $\{\arg a_1(t,\xi)\}_\Gamma = 0$ for sufficiently large ξ which implies (5.4.8). If q is odd, then

$$a\xi + b = e^{(q-1)i\vartheta}a_1,\ a_1 = |t-t_0|^{q-1}(t-t_0)\xi + b_1,\ b_1 = be^{(1-q)i\vartheta}.$$

Since $x(t-t_0)^{-1}b_1|_{x=0} = |t'(0)|^{(q-1)/2}\mu$, one can easily deduce that $(2\pi)^{-1}\{\arg a_1(t,\xi)\}_\Gamma$ is equal to 1 when $\mathrm{Im}\,\mu < 0$ and equal to 0 when $\mathrm{Im}\,\mu > 0$ for sufficiently large ξ. Thus we obtain (5.4.8)again. □

<u>Proof of Theorem 5.4.1.</u> By Remark 5.3.8, conditions (5.3.17) and (5.4.4) are obviously equivalent. Using (5.4.1) and the relations

$$\mu_m^{\pm}(k) = (t-t_m^{\pm})^{q_m^{\pm}} a_{l-1}^k / a_l^{\pm}\Big|_{t=t_m^{\pm},\ \xi=\pm 1} (t'(x_m^{\pm}))^{-q_m^{\pm}}$$

(see (5.4.2)) and applying successively Lemmas 5.4.2 and 5.4.3, one deduces that

$$\{\arg a^k(t,\pm\xi)\}_\Gamma = \{\arg \tilde{a}_l^{\pm}\}_\Gamma + 2\pi \sum_{1\le m\le n^{\pm}} \eta_m^{\pm}(k)$$

for sufficiently large ξ which proves formula (5.4.5). □

<u>5.4.2.</u> We now investigate the regularity of solutions to degenerate pseudodifferential equations. Let $\Sigma_1^{\pm} = \{t_1^{\pm}, t_2^{\pm}, \dots\}$ be the set of all points on Γ where $a_l^{\pm}$ vanishes exactly to first order, and define the numbers $\mu_m^{\pm}(k)$ as in (5.4.2):

$$\mu_m^{\pm}(k) = a_{l-1}^{\pm}(x)(x-x_m^{\pm})/a_l^{\pm}(x)\big|_{x_m^{\pm}} \mp i(k-1/2),\ t_m^{\pm} = t(x_m^{\pm}).$$

<u>Theorem 5.4.4.</u> Let $k_1 < k_2$, and assume (5.3.3) for $k=k_1$ and $\mathrm{Im}\,\mu_m^{\pm}(k) \ne 0$ for $k \in (k_1,k_2]$ and all $t_m^{\pm} \in \Sigma_1^{\pm}$. Then $u \in W_2^{k_1+l-1}$ and $Au \in W_2^{k_2}$ imply $u \in W_2^{k_2+l-1}$.

<u>Proof.</u> By the first assumption, there exist neighborhoods $U^{\pm}$ of $\Sigma^{\pm} \setminus \Sigma_1^{\pm}$ such that $\mathrm{Re}\,\varrho^{\pm} a_{l-1}^k(t,\pm 1) > 0$ on $\bar{U}^{\pm}$ for all $k \in [k_1,k_2]$ since we may assume $|Da_l^{\pm}|$ to be arbitrarily small on $\bar{U}^{\pm}$. On the other hand, the sets $M^{\pm} = (\Gamma \setminus \bar{U}^{\pm}) \cap \Sigma^{\pm} = (\Gamma \setminus \bar{U}^{\pm}) \cap \Sigma_1^{\pm}$ are finite and using the assumptions, we observe that $\mathrm{Re}\,\varrho^{\pm} a_{l-1}^k(t,\pm 1) > 0$ on $M^{\pm}$ for any $k \in [k_1,k_2]$. Therefore condition (5.3.3) is satisfied with the same functions $\varrho^{\pm}$ when $k \in [k_1,k_2]$. Consequently, by formulas (5.3.4) and (5.3.6), the

indices of (5.3.1) coincide for $k \in [k_1,k_2]$. Since $W_2^{k_2}$ is dense in $W_2^{k_1}$, we get the conclusion as in Theorem 1.5.1. □

Corollary 5.4.5. Let $\Sigma_1^{\pm} = \emptyset$ and suppose there exist functions $\rho^{\pm} \in C^{\infty}$ such that $\rho^{\pm}(t) \neq 0$ $(t \in \Gamma)$ and

$$\operatorname{Re} \rho^{\pm} a_1^{\pm} \geq 0 \text{ on } \Gamma, \quad \operatorname{Re} \rho^{\pm} a_{1-1}^{\pm} > 0 \text{ on } \Sigma^{\pm}. \tag{5.4.9}$$

Then A is hypoelliptic, i.e. $u \in \mathcal{D}'$ and $Au \in C^{\infty}$ imply $u \in C^{\infty}$. Moreover, $A \in L(C^{\infty})$ is a Fredholm operator with index (5.3.19).

This is an immediate consequence of Theorem 5.4.4 and Corollary 5.3.1o. In particular, (5.4.9) holds if condition (5.3.18) holds and $a_1^{\pm}$ satisfy condition (R) (see the proof of Theorem 5.3.6). As a special case of Theorem 5.4.4, from Theorem 5.4.1 we obtain

Corollary 5.4.6. Under the assumptions (5.4.1), (5.4.4) for $k=k_1$ and $\operatorname{Im} \mu_m^{\pm}(k) \neq 0$ for all $t_m^{\pm} \in \Sigma_1^{\pm}$ and $k \in (k_1,k_2]$, $u \in W_2^{k_1+1-1}$ and $Au \in W_2^{k_2}$ imply $u \in W_2^{k_2+1-1}$.

Corollary 5.4.7. Assume (5.4.1), $q_m^{\pm} = 1$ $(m=1,\dots,m^{\pm} \leq n^{\pm})$ and (5.4.4) for $m = m^{\pm}+1,\dots,n^{\pm}$. Then $A \in L(C^{\infty})$ is a Fredholm operator with index

$$\operatorname{ind} A = (2\pi)^{-1} \{\arg \tilde{a}_1^{-}/\tilde{a}_1^{+}\}_{\Gamma} - \sum_{m^+ < m \leq n^+} \eta_m^{+} + \sum_{m^- < m \leq n^-} \eta_m^{-} - m^{+} .$$

Corollary 5.4.7 is a consequence of Theorem 5.4.1, since $\operatorname{Im} \mu_m^{+}(k) < 0$ $(m=1,\dots,m^{+})$ and $\operatorname{Im} \mu_m^{-}(k) > 0$ $(m=1,\dots,m^{-})$ for all sufficiently large k and C^{∞} is the projective limit of the spaces W_2^k. Note that A is always a Fredholm operator in C^{∞} with index $(2\pi)^{-1} \{\arg \tilde{a}_1^{-}/\tilde{a}_1^{+}\}_{\Gamma} - n^{+}$ when $a_1^{\pm}$ have only zeros of first order on Γ.

5.4.3. We finally give two examples.

Example 5.4.8. Consider the differential operator $A = a(x)D+1$, where $a \in C^{\infty}$ is such that

$$a(0) = 0, \; a(x) = e^{-1/x^2} \sin(\pi/x), \; x \in [-2/3,3/2] \setminus \{0\};$$

$$a(x) = 1, \; x \in [-\pi,\pi] \setminus (-1,2); \; a(x) > 0, \; x \in (-1,-2/3) \cup (3/2,2).$$

a(x) satisfies condition (R), vanishes to infinite order at the origin and has zeros of first order at $x_n = 1/n$, $n=1,\pm 2,\pm 3,\dots$. A has the symbol $ia(x)\xi +1$ so that

$, a^k = ia(x)\xi + a_o^k\ ,\ a_o^k = 1 + (k-1/2)(Da)(x)$ (see (5.3.2)).

Since, for given $k \in \mathbb{R}$, $|(k-1/2)Da| < 1$ when $|x| \leq 1/n$ and n is sufficiently large, we see that condition (5.3.17) holds if and only if

$$a_o^k(1/n) = 1+(k-1/2)(-1)^n n^2 \pi e^{-n^2} \neq 0,\ n=1,\pm 2,\pm 3,\dots . \qquad (5.4.10)$$

If (5.4.10) is satisfied, then by formula (5.3.4) and Lemmas 5.4.2 and 5.4.3 one obtains that A is a Fredholm operator in W_2^k with index $-2z(k)$, where $z(k)$ denotes the number of odd indices $n \in \mathbb{Z}\setminus\{-1\}$ satisfying $e^{n^2}/n^2\pi < k-1/2$. Since $z(k) \to \infty$ as $k \to \infty$, we observe that A cannot be a Fredholm operator in C^∞.

Example 5.4.9. Consider the operator

$$A = aI + bS_\Gamma + T^+ + T^- + K\ , \qquad (5.4.11)$$

where $a,b \in C^\infty$, S_Γ is the Cauchy integral operator on Γ, K a smoothing operator and $T^\pm$ as in (5.1.20). Setting $a^\pm = (a\pm b)/2$ and using (5.1.22) and Example 5.1.4, we obtain that A is a PDO of order 0 with symbol $a_0+a_{-1}+\dots$, where

$$a_0 = a^+h(\xi)+a^-h(-\xi),\ a_{-1} = it'\{k^+(t,t)h(\xi)+k^-(t,t)h(-\xi)\}|\xi|^{-1}.$$

Hence

$$a_{-1}^k(t,\xi) = \{k^+(t,t)t' - (k-1/2)Da^+\}ih(\xi)/\xi$$
$$- \{k^-(t,t)t'+(k-1/2)Da^-\}ih(-\xi)/\xi;$$

see (5.3.2). Theorem 5.3.6 then yields the following result.

Theorem 5.4.10. Let A be the operator (5.4.11) and suppose $a^\pm$ satisfy condition (R). The closed operator

$$A : W_2^{k-1} \to W_2^k,\ D_k(A) = \{u \in W_2^{k-1} : Au \in W_2^k\} \qquad (5.4.12)$$

is a Fredholm operator if and only if

$$\varliminf_{\xi\to\infty} \inf_{t\in\Gamma} |a^\pm\xi + it'k^\pm(t,t) \mp i(k-1/2)Da^\pm| > 0\ . \qquad (5.4.13)$$

If (5.4.13) is satisfied, then (5.4.12) has the index

$$\begin{aligned} \operatorname{ind} A = {} & (2\pi)^{-1}\{\arg[a^-\xi+it'k^-(t,t)+i(k-1/2)Da^-]\}_\Gamma \\ & - (2\pi)^{-1}\{\arg[a^+\xi+it'k^+(t,t)-i(k-1/2)Da^+]\}_\Gamma \end{aligned} \qquad (5.4.14)$$

for sufficiently large ξ.

Note that in view of Lemma 5.3.2 the domain of definition of (5.4.12) can be written $D_k(A) = \{ u \in W_2^{k-1} : a^{\pm}P^{\pm}u \in W_2^k \}$. Finally, we consider the case when $a^{\pm}$ take the form

$$a^{\pm} = \beta^{\pm}\tilde{a}^{\pm}, \quad \tilde{a}^{\pm}(t) \neq 0 \quad (t \in \Gamma) , \tag{5.4.15}$$

with $\beta^{\pm}$ defined in (5.4.1). Concerning the numbers (5.4.2) we have

$$\begin{aligned} &\mu_m^{\pm} = \mu_m^{\pm}(k) = it'(x_m^{\pm})k^{\pm}(t_m^{\pm},t_m^{\pm})/(Da^{\pm})(x_m^{\pm}) \mp i(k-1/2) , \\ &m = 1,\ldots,m^{\pm} , \\ &\mu_m^{\pm} = it'(x_m^{\pm})k^{\pm}(t_m^{\pm},t_m^{\pm})(q_m^{\pm})!/(D^{q_m^{\pm}}a^{\pm})(x_m^{\pm}) , \\ &m = m^{\pm}+1,\ldots,n^{\pm} , \end{aligned} \tag{5.4.16}$$

where $q_m^{\pm}=1$, $m=1,\ldots,m^{\pm}$, and $q_m^{\pm}>1$, $m = m^{\pm}+1,\ldots,n^{\pm}$. Defining $\eta_m^{\pm}(k)$ as in (5.4.3), from Theorem 5.4.1 and Corollary 5.4.7 we obtain

Corollary 5.4.11. Assume (5.4.15).

(i) The operator (5.4.12) is Fredholm if and only if condition (5.4.4) is satisfied with $\mu_m^{\pm}$ defined in (5.4.16), and under this hypothesis,

$$\text{ind } A = (2\pi)^{-1}\{\arg \tilde{a}^-/\tilde{a}^+\}_{\Gamma} - \sum_{1\le m\le n^+} \eta_m^+(k) + \sum_{1\le m\le n^-} \eta_m^-(k) .$$

(ii) If (5.4.4) is valid for $m = m^{\pm}+1,\ldots,n^{\pm}$, then $A \in L(C^{\infty})$ is a Fredholm operator with index

$$(2\pi)^{-1}\{\arg \tilde{a}^-/\tilde{a}^+\}_{\Gamma} - \sum_{m^+< m\le n^+} \eta_m^+ + \sum_{m^-< m\le n^-} \eta_m^- - m^+ .$$

In particular, if $a^{\pm}$ have only zeros of first order, then (5.4.10) is always a Fredholm operator in C^{∞} with index $(2\pi)^{-1}\{\arg \tilde{a}^-/\tilde{a}^+\}_{\Gamma} - n^+$.

5.5. On the degenerate oblique derivative problem in the plane

5.5.1. Let $\Omega \subset \mathbb{R}^2$ be a bounded domain such that its boundary Γ is a simple closed C^{∞} curve. The oblique derivative or Poincaré problem is to find a real-valued function u satisfying

$$\begin{aligned} &(\Delta u)(x) = 0 , \quad x = (x_1,x_2) \in \Omega , \\ &(Tu)(x) = a\,\partial u/\partial n + b\,\partial u/\partial \vartheta + cu|_{\Gamma} = f(x), \quad x \in \Gamma , \end{aligned} \tag{5.5.1}$$

where a,b,c and f are given real functions on Γ and Δ denotes the Laplacian in $\mathbb{R}^2$. By ϑ and n we denote the arc coordinate on Γ and the inner normal to Γ, respectively. Since a conformal mapping does not change the Laplacian and the form of the boundary condition in (5.5.1), we may assume that Ω is the unit disk $x_1^2 + x_2^2 < 1$.

Throughout this section we suppose that $a,b,c \in C^{\infty}(\Gamma)$ and that all occurring function spaces consist of real-valued functions. Furthermore, let $\tilde{C}^{\infty}(\bar{\Omega})$ be the closed subspace $C^{\infty}(\bar{\Omega}) \cap H(\Omega)$ of $C^{\infty}(\bar{\Omega})$ and $\tilde{W}_2^k(\Omega)$ the completion of $\tilde{C}^{\infty}(\bar{\Omega})$ with respect to the norm in $W_2^k(\Omega)$, where $k \geq 0$ and $H(\Omega)$ denotes the space of harmonic functions in Ω.

We now reduce problem (5.5.1) to a pseudodifferential equation on Γ. We shall identify $(x_1,x_2) \in \bar{\Omega}$ with $t = x_1 + ix_2 = \varrho e^{i\vartheta}$ and $(\tau_1,\tau_2) \in \Gamma$ with $\tau = \tau_1 + i\tau_2 = e^{i\omega}$ in the following. Consider the Neumann problem

$$(\Delta u)(x) = 0 \ , \ x \in \Omega \ , \quad \partial u/\partial n|_{\Gamma} = g(x) \ , \ x \in \Gamma \ . \tag{5.5.2}$$

It is well-known (Hörmander [2, Chap. 10], Michlin [1, Chap. 15]) that for any $g \in W_2^{k+1/2}(\Gamma)$ $(k \geq 0)$ satisfying $\int_{-\pi}^{\pi} g(\vartheta)d\vartheta = 0$, (5.5.2) has the solution

$$u(t) = (Pg)(t) = \pi^{-1} \int_{\Gamma} g(\tau) \ln r^{-1} d\omega, \ r = |\tau - t|, \ t \in \Omega \tag{5.5.3}$$

in $\tilde{W}_2^{k+2}(\Omega)$ which is unique modulo constants. Consequently,

$$P \in L(W_2^{k+1/2}(\Gamma), \tilde{W}_2^{k+2}(\Omega)), \ k \geq 0, \ P \in L(C^{\infty}(\Gamma), \tilde{C}^{\infty}(\bar{\Omega}))$$

are Fredholm operators with index 0. We show that $A = TP$ is a PDO of the form (5.4.11) on Γ.

Let $g \in C^{\infty}(\Gamma)$. (5.5.3) is the real part of the function

$$F(t) = \pi^{-1} \int_{\Gamma} g(\tau) \ln(\tau - t)^{-1} d\omega \ , \ t \in \Omega \tag{5.5.4}$$

which is holomorphic in Ω. Setting $F(t) = u(x_1,x_2) + iv(x_1,x_2)$, we differentiate (5.5.4) with respect to x_1. By the Cauchy-Riemann equation $\partial v/\partial x_1 = -\partial u/\partial x_2$, we have

$$\begin{aligned} \partial u/\partial x_1 - i\partial u/\partial x_2 &= \pi^{-1} \int_{\Gamma} g(\tau)(\tau - t)^{-1} d\omega \\ &= (\pi i)^{-1} \int_{\Gamma} g(\tau)\tau^{-1}(\tau - t)^{-1} d\tau, \ t \in \Omega \ . \end{aligned} \tag{5.5.5}$$

Letting t tend to a point on the boundary in (5.5.5) and using the Plemel'-Sochocki formulas (cf. Muschelischwili [1, § 16]), one obtains

$$\partial u/\partial x_1 - i\,\partial u/\partial x_2 = g(t)t^{-1} + \pi^{-1}\int_\Gamma g(\tau)(\tau - t)^{-1}d\omega,\quad t\in\Gamma\ .$$

Setting $\tau - t = re^{i\gamma}$ with $\tau = e^{i\omega}$ and $t = e^{i\vartheta}$ and taking real and imaginary parts, the last equality gives

$$\partial u/\partial x_1 = g(t)\cos\vartheta + \pi^{-1}\int_\Gamma g(\tau)r^{-1}\cos\gamma\, d\omega,$$

$$\partial u/\partial x_2 = g(t)\sin\vartheta + \pi^{-1}\int_\Gamma g(\tau)r^{-1}\sin\gamma\, d\omega\ .$$

Furthermore, using the relations

$$\partial u/\partial\vartheta = \cos\theta\ \partial u/\partial x_1 + \sin\theta\ \partial u/\partial x_2\ ,$$

$$\partial u/\partial n = -\sin\theta\ \partial u/\partial x_1 + \cos\theta\ \partial u/\partial x_2,\quad \theta = \vartheta + \pi/2\ ,$$

$$\cos\gamma\, d\omega/r = -\sin\vartheta\, d\tau/(\tau - t)\ ,$$

$$\sin\gamma\, d\omega/r = \cos\vartheta\, d\tau/(\tau - t) - d\tau/2\pi\tau t\ ,$$

we finally obtain

$$\partial u/\partial n = -g - i(2\pi)^{-1}\int_\Gamma g(\tau)\tau^{-1}d\tau\ ,$$

$$\partial u/\partial\vartheta = \pi^{-1}\int_\Gamma g(\tau)(\tau - t)^{-1}d\tau - (2\pi)^{-1}\int_\Gamma g(\tau)\tau^{-1}d\tau\ .$$

Hence

$$A = TP = -aI + ibS_\Gamma + cV + K\ , \tag{5.5.6}$$

where S_Γ is the Cauchy integral operator on Γ, K a smoothing operator and

$$(Vg)(t) = \pi^{-1}\int_\Gamma g(\tau)\ln|\tau - t|^{-1}d\omega\ .$$

Using the notation in (5.1.2o), we can obviously write

$$Vg = (2\pi t')^{-1}\int_\Gamma \{\ln(1-t/\tau) + \ln(1-\tau/t)\}\, g(\tau)d\tau\ ,\quad t' = ie^{i\vartheta}.$$

Therefore, by (5.1.22), V has the principal symbol $|\xi|^{-1}$. Consequently, by virtue of (5.5.6), A is a PDO of order 0 on Γ with symbol $a_0 + a_{-1} + \dots$, where

$$a_0 = -a + ib\xi/|\xi|\ ,\quad a_{-1} = c/|\xi|\ . \tag{5.5.7}$$

5.5.2. Next, we recall the classical index formula for problem (5.5.1) in the elliptic case. (5.5.1) generates a continuous linear operator

$$T : \widetilde{W}_2^{k+2}(\Omega) \to W_2^{k+1/2}(\Gamma)$$

for any $k \geq 0$ (cf. Lions and Magenes [1, Chap. 2]).

Theorem 5.5.1. $T \in L(\tilde{W}_2^{k+2}(\Omega), W_2^{k+1/2}(\Gamma))$, $k \geq 0$, is a Fredholm operator if and only if

$$M(t) = a(t) + ib(t) \neq 0 \ , \ t \in \Gamma \ . \tag{5.5.8}$$

If (5.5.8) is satisfied, then T has the index $\pi^{-1}\{\arg M\}_\Gamma$.

<u>Proof.</u> Since $P \in L(W_2^{k+1/2}(\Gamma), \tilde{W}_2^{k+2}(\Omega))$ is a Fredholm operator with index 0, $T \in L(\tilde{W}_2^{k+2}(\Omega), W_2^{k+1/2}(\Gamma))$ is Fredholm if and only if so is $A = TP \in L(W_2^{k+1/2}(\Gamma))$, and then ind T = ind A. Now Theorem 5.5.1 is a consequence of (5.5.7) and Theorem 5.1.14. □

When M has zeros on Γ we consider the closed operator

$$\begin{aligned} &T : \tilde{W}_2^{k+2}(\Omega) \to W_2^{k+3/2}(\Gamma) \ , \ k \geq 0 \ , \\ &D(T) = \{ u \in \tilde{W}_2^{k+2}(\Omega) \ : \ Tu \in W_2^{k+3/2}(\Gamma) \} \ . \end{aligned} \tag{5.5.9}$$

Let $\Sigma = \{t \in \Gamma : M(t) = 0\}$ and set $b^k = M\xi - c + i(k+1)D_\vartheta M$.

Theorem 5.5.2. Suppose M satisfies condition (R) on Γ (see 5.3.2). (5.5.9) is a Fredholm operator if and only if there exists $\xi_0 > 0$ such that

$$|b^k(t,\xi)| \geq C > 0 \ , \ t \in \Gamma \ , \ \xi \geq \xi_0 \ . \tag{5.5.10}$$

If (5.5.10) holds, then (5.5.9) has the index

$$\operatorname{ind} T = \pi^{-1}\{\arg b^k(t,\xi)\}_\Gamma \ , \ \xi \geq \xi_0 \ . \tag{5.5.11}$$

<u>Proof.</u> As in Theorem 5.5.1, we observe that (5.5.9) is a Fredholm operator if and only if the closed operator

$$A = TP : W_2^{k+1/2}(\Gamma) \to W_2^{k+3/2}(\Gamma), \ D(A) = \{ u \in W_2^{k+1/2} : Au \in W_2^{k+3/2} \}$$

is Fredholm, and their indices coincide. By (5.5.7), the function (5.3.2) corresponding to A can be written

$$\begin{aligned} a^{k+3/2} &= a_0 + a_{-1} + (k+1)(i\xi)^{-1} D_\vartheta a_0 = -\bar{M} + (c + i(k+1)D_\vartheta \bar{M})|\xi|^{-1} \\ &= -\overline{b^k}\,|\xi|^{-1} \text{ when } \xi > 0, \ a^{k+3/2} = -b^k\,|\xi|^{-1} \text{ when } \xi < 0 \ . \end{aligned}$$

Now the theorem follows from Theorem 5.3.6. □

Analogously, setting $\alpha = M\xi - c$, from Corollary 5.3.10 we obtain

Corollary 5.5.3. If M satisfies condition (R) and has only zeros of order ≥ 2, condition (5.5.10) is equivalent to

$$\varliminf_{\xi\to\infty} \inf_{t\in\Gamma} |\alpha(t,\xi)| > 0 , \qquad (5.5.12)$$

and for any $k \geq 0$, the index formula (5.5.11) can then be replaced by

$$\operatorname{ind} T = \pi^{-1} \{\arg \alpha(t,\xi)\}_\Gamma \qquad (5.5.13)$$

where ξ is sufficiently large. Moreover, under hypothesis (5.5.12), $T \in L(\tilde{C}^\infty(\bar{\Omega}), C^\infty(\Gamma))$ is a Fredholm operator with index (5.5.13).

We now consider the special case when M has only isolated zeros of finite order. Assume that M takes the form

$$\begin{aligned} &M(t) = \beta(t)\tilde{M}(t) , \quad \beta(t) = \prod_{1\leq j\leq n} (t-t_j)^{q_j}, \quad q_j \in \mathbb{N} , \\ &\tilde{M}(t) \neq 0 \ (t\in\Gamma) , \end{aligned} \qquad (5.5.14)$$

where $t_j = t(\vartheta_j)$, $-\pi\leq\vartheta_j<\pi$, are distinct points on Γ. Let $q_j = 1$ $(j=1,\dots,m\leq n)$ and define the numbers

$$\mu_j = \mu_j(k) = -c(\vartheta_j)/(D_\vartheta M)(\vartheta_j) + i(k+1) , \quad j\leq m;$$

$$\mu_j = -c(\vartheta_j)q_j!/(D_\vartheta^{q_j}M)(\vartheta_j) \quad , \ j> m;$$

$$\eta_j = \eta_j(k) = (q_j\pm1)/2 \text{ if } q_j \text{ is odd and } \pm \operatorname{Im}\mu_j< 0 ,$$

$$\eta_j = q_j/2 \text{ if } q_j \text{ is even.}$$

Using Lemmas 5.4.2 and 5.4.3, from Theorem 5.5.2 and Corollary 5.5.3 we then obtain

<u>Theorem 5.5.4.</u> Assume (5.5.14). Then (5.5.9) is a Fredholm operator if and only if, for $j=1,\dots,m$,

$$\operatorname{Im}\mu_j \neq 0 \text{ when } q_j \text{ is odd, } \operatorname{Im}\mu_j \neq 0 \text{ or } \operatorname{Re}\mu_j> 0 \text{ when } q_j \text{ is even} , \qquad (5.5.15)$$

and formula (5.5.11) can be written

$$\operatorname{ind} T = \pi^{-1} \{\arg \tilde{M}\}_\Gamma + 2 \sum_{1\leq j\leq n} \eta_j .$$

Furthermore, if (5.5.15) is satisfied for $j=m+1,\dots,n$, then $T\in L(\tilde{C}^\infty(\bar{\Omega}), C^\infty(\Gamma))$ has the index

$$\pi^{-1} \{\arg \tilde{M}\}_\Gamma + 2(\eta_{m+1}+\dots+\eta_n).$$

<u>5.5.3.</u> Finally, we derive an existence and uniqueness result for the degenerate oblique derivative problem.

Theorem 5.5.5. Assume $\Sigma \neq \emptyset$, $k \geq 2$ and

$$a \leq 0, \ c \geq 0 \text{ on } \Gamma, \tag{5.5.16}$$

$$c > 0 \text{ on } \{t \in \Gamma : a(t) = 0\} \tag{5.5.17}$$

$$c + (k+1)D_{\vartheta} b > 0 \text{ on } \Sigma . \tag{5.5.18}$$

Then the operator (5.5.9) is invertible.

Proof. First, we verify that (5.5.9) has index 0. As in the proof of Theorem 5.5.2, it is sufficient to show that $A = TP$ as a closed map of $W_2^{k+1/2}(\Gamma)$ into $W_2^{k+3/2}(\Gamma)$ has index 0. By (5.5.7), (5.5.16) and (5.5.18)

$$\operatorname{Re} a_0 \geq 0 \text{ on } \Gamma \times \{\pm 1\}, \ \operatorname{Re} \{a_{-1} + (k+1)(i\xi)^{-1} D_{\vartheta} a_0\}$$

$$= c + (k+1) D_{\vartheta} b > 0 \quad \text{on } \Sigma \times \{\pm 1\}$$

so that Theorem 5.3.3 implies the assertion. It remains to prove that the homogeneous problem

$$\Delta u = 0, \ Tu = a\,\partial u/\partial n + b\,\partial u/\partial \vartheta + cu|_{\Gamma} = 0 \tag{5.5.19}$$

has only the trivial solution in $\tilde{W}_2^{k+2}(\Omega)$. Arguing by contradiction, we suppose (5.5.19) has a solution $u \neq 0$ in $\tilde{W}_2^{k+2}(\Omega)$. Since $k \geq 2$, u is twice continuously differentiable in $\bar{\Omega}$, and by virtue of (5.5.17) and $\Sigma \neq \emptyset$, u is not equal to a constant. Let $x_0 \in \bar{\Omega}$ be a positive maximum or negative minimum of u. Then by the maximum principle for the Laplacian in $\mathbb{R}^2$ (cf. e.g. Miranda [1]),

$$x_0 \in \Gamma, \ \partial u/\partial \vartheta|_{x_0} = 0, \ \partial u/\partial n|_{x_0} < 0 \ (>0)$$

in the case of a positive maximum (negative minimum). Together with (5.5.16) and (5.5.17), this implies $Tu|_{x_0} \neq 0$, hence a contradiction. □

Note that condition (5.5.18) can be dropped in Theorem 5.5.5 when b vanishes only to order ≥ 2 on Σ.

5.6. A class of pseudodifferential operators degenerating at one point

5.6.1. In this section we study Fredholm property and index for a class of PDOs whose principal symbol vanishes to finite order at one point on the curve. In contrast to Theorem 5.4.1, the index may depend on an arbitrary (finite) number of lower order terms of the symbol. Let A be a classical PDO of order l on Γ with symbol $a_l+a_{l-1}+\dots$, and set $a^{\pm}_{l-j}(t) = a_{l-j}(t,\pm 1)$, $j \in \mathbb{N}_o$. We assume that the functions $a^{\pm}_l$ take the form

$$a^{\pm}_l = (t-t_o)^{q^{\pm}}\tilde{a}^{\pm}_l \ , \ q^{\pm} \in \mathbb{N}_o \ , \ t_o = t(0)\in\Gamma \ , \ \tilde{a}^{\pm}_l(t)\neq 0 \ (t\in\Gamma). \quad (5.6.1)$$

Let us define natural numbers $j^{\pm}$ $(0\le j^{\pm}\le q^{\pm})$ by the relations

$$\begin{aligned} &l-j-v(a^{\pm}_{l-j}) \le l-j^{\pm}-v(a^{\pm}_{l-j^{\pm}}) \quad \text{for } j > j^{\pm} \ , \\ &l-j-v(a^{\pm}_{l-j}) < l-j^{\pm}-v(a^{\pm}_{l-j^{\pm}}) \quad \text{for } j < j^{\pm} \ , \end{aligned} \quad (5.6.2)$$

where $v(a)$ denotes the order of the zero of a function $a\in C^{\infty}$ at the point $x=0$. Furthermore, we set

$$l^{\pm} = j^{\pm} + v(a^{\pm}_{l-j^{\pm}}) \ , \ r^{\pm} = l^{\pm} - j^{\pm} \ . \quad (5.6.3)$$

By (5.6.2), $l^{\pm}\le q^{\pm}$. Putting $b^{\pm}_j = (\mp i)^j a^{\pm}_{l-l^{\pm}+j}$, with A we associate the differential operators

$$B^{\pm} = \sum_{0\le j\le l^{\pm}} b^{\pm}_j D^j \quad (5.6.4)$$

with periodic coefficients on $[-\pi,\pi]$, which have a singularity at $x = 0$ when $q^{\pm}> 0$. By Theorem 5.1.11, we obtain the equality

$$A = (B^+P^++B^-P^-)(\Lambda^{l-l^+}P^++\Lambda^{l-l^-}P^-) + T^+P^++T^-P^- \ , \quad (5.6.5)$$

where $T^{\pm}\in OPCS^{l-l^{\pm}-1}$ and Λ^k $(k \in \mathbb{R})$ and $P^{\pm}$ are defined by (5.1.7) and (5.1.14), respectively. In view of (5.6.2), (5.6.3) and Corollary 2.1.5, exactly $r^{\pm}$ characteristic factors of $B^{\pm}$ (at the origin) vanish identically and $\varkappa(B^{\pm}) = 0$ (cf. 2.4.2 and (2.1.4)). Furthermore, the characteristic equation of $B^{\pm}$ is

$$x^{-\mu}B^{\pm}(x^{\mu})|_{x=0} = 0 \ ; \quad (5.6.6)$$

see (2.1.12). Let $\mu^{\pm}_j$ $(j=1,\dots,r^{\pm})$ be the roots of (5.6.6). Moreover, let $\mu^{\pm}_j x^{1-q^{\pm}_j}$ be the lowest terms of the non-vanishing characteristic factors of $B^{\pm}$, where

$$1< q^{\pm}_j \le q^{\pm}_{j+1} \ , \ j=r^{\pm}+1,\dots,l^{\pm}-1 \ . \quad (5.6.7)$$

The numbers $\mu^{\pm}_j$ and $q^{\pm}_j$ $(j> r^{\pm})$ may be computed using Lemma 2.2.1. Henceforth we shall assume that

$$q_j^\pm \in \mathbb{N}\ ,\ j = r^\pm+1,\ldots,l^\pm\ . \tag{5.6.8}$$

Define $q^\pm(j) = q_o^\pm+\ldots+q_j^\pm$, where $q_o^\pm = 0$, $q_j^\pm = 1$ $(j=1,\ldots,r^\pm)$. In analogy to Chap. 2.6 we introduce the weighted Sobolev spaces

$$W^k_{2,\pm} = W^k_{2,B^\pm} = \{u\in \mathcal{D}' : (t-t_o)^{q^\pm(j)}D^j u\in W_2^k,\ j = 0,\ldots,l^\pm\},$$

$$W^k_{2,A} = \{u\in\mathcal{D}' : \Lambda^{l-l^+}P^+u\in W^k_{2,+},\ \Lambda^{l-l^-}P^-u\in W^k_{2,-}\}$$

equipped with the canonical norms. By (5.6.5), (5.6.8) and the definition of $q^\pm(j)$, $A\in L(W^k_{2,A},W^k_2)$ for any $k\in\mathbb{R}$. In order to investigate the Fredholm property of this operator, we introduce the conditions

$$\operatorname{Re}\mu_j^\pm \neq -1/2+k,\ j = 1,\ldots,r^\pm\ ; \tag{5.6.9}$$

$$\operatorname{Re}\mu_j^\pm \neq 0 \text{ if } q_j^\pm \text{ is odd, } \operatorname{Re}\mu_j^\pm \neq 0 \text{ or } \operatorname{Im}(\pm\mu_j^\pm) < 0 \text{ if } q_j^\pm \text{ is even, } j = r^\pm+1,\ldots,l^\pm\ . \tag{5.6.10}$$

Further, let $\eta^\pm(k)$ denote the number of the $\mu_j^\pm(j=1,\ldots,r^\pm)$ satisfying $\pm\operatorname{Re}(\mu_j^\pm+1/2-k)< 0$ and let

$$\eta^\pm = \sum_{r^\pm< j\le l^\pm} \eta_j^\pm\ ,$$

where $\eta_j^\pm = q_j^\pm/2$ for even $q_j^\pm$, $\eta_j^\pm = (q_j^\pm+1)/2$ for odd $q_j^\pm$ and $\pm\operatorname{Re}\mu_j^\pm<0$, $\eta_j^\pm = (q_j^\pm-1)/2$ for odd $q_j^\pm$ and $\pm\operatorname{Re}\mu_j^\pm>0$.

<u>Theorem 5.6.1.</u> Assume (5.6.1), (5.6.8), (5.6.9) and (5.6.10). Then $A\in L(W^k_{2,A},W^k_2)$ is a Fredholm operator with index

$$\operatorname{ind} A = (2\pi)^{-1}\{\arg \tilde a_l^-/\tilde a_l^+\}_\Gamma - \eta^+(k)-\eta^+ + \eta^-(k)+\eta^-. \tag{5.6.11}$$

Theorem 5.4.1 shows that Theorem 5.6.1 does not hold, in general, if one of the conditions (5.6.9) and (5.6.10) is violated. Since $\eta^+(k) = r^+$, $\eta^-(k) = 0$ for sufficiently large k and C^∞ is the projective limit of the spaces $W^k_{2,A}$ and W^k_2, as a consequence of Theorem 5.6.1 we obtain

<u>Corollary 5.6.2.</u> Under hypotheses (5.6.1), (5.6.8) and (5.6.10), $A\in L(C^\infty)$ is a Fredholm operator with index

$$(2\pi)^{-1}\{\arg \tilde a_l^-/\tilde a_l^+\}_\Gamma - r^+ - \eta^+ + \eta^-\ .$$

<u>Proof of Theorem 5.6.1.</u> Set $B = B^+P^+ + B^-P^-$ and define the space

$$W^k_{2,B} = \{ u \in \mathcal{D}' : P^+u \in W^k_{2,+},\ P^-u \in W^k_{2,-} \}$$

endowed with the canonical norm. By (5.6.5) and the facts that $\Lambda^{l-l^+}P^+ + \Lambda^{l-l^-}P^-$ is an isomorphism of $W^k_{2,A}$ onto $W^k_{2,B}$ and that

$$T^+P^+ + T^-P^- : W^k_{2,A} \to W^k_2$$

is compact, it suffices to show that, under the hypotheses of the theorem, $B : W^k_{2,B} \to W^k_2$ has index (5.6.11). Furthermore,

$$B = C^+C^- + K,\quad C^+ = B^+P^+ + P^-,\quad C^- = P^+ + B^-P^- \tag{5.6.12}$$

where K is some smoothing operator. Therefore it is sufficient to prove that $C^{\pm} \in L(W^k_{2,C^{\pm}}, W^k_2)$ are Fredholm with indices

$$\mp(2\pi)^{-1} \{\arg \tilde{a}^{\pm}_l\}_{\Gamma} \mp \eta^{\pm}(k) \mp \eta^{\pm}, \tag{5.6.13}$$

where

$$W^k_{2,C^{\pm}} = \{ u \in \mathcal{D}' : P^{\mp}u \in W^k_2,\ P^{\pm}u \in W^k_{2,\pm} \}$$

(equipped with the canonical norm). Consider C^+, for example. Setting

$$C_j = [(t-t_o)^{q^+_j} D - c_j]\, P^+ + P^-,\quad c_j = \mu^+_j(t'(0))^{q^+_j},\quad j=1,\dots,l^+ ,$$

we introduce the closed operators

$$C_j : W^k_2 \to W^k_2,\quad D(C_j) = \{ u \in W^k_2 : (t-t_o)^{q^+_j} DP^+u \in W^k_2 \}. \tag{5.6.14}$$

By Lemma 5.3.1, they are densely defined in W^k_2 and using (5.6.7), it is easily seen that

$$D(C_{l^+} \cdots C_1) = W^k_{2,C^+}.$$

Applying Theorem 5.4.1 to the operators (5.6.14), Theorem 5.1.14 to $\tilde{C} = \tilde{a}^+_l P^+ + P^- \in L(W^k_2)$ and the well-known index formula for a product of densely defined Fredholm operators (cf. Goldberg [1, Th. IV.2.7]), we obtain that

$$C = \tilde{C}\, C_{l^+} \cdots C_1 \in L(W^k_{2,C^+}, W^k_2)$$

is Fredholm with index

$$-\eta^+(k) - \eta^+ - (2\pi)^{-1} \{\arg \tilde{a}^+_l\}_{\Gamma} .$$

To show (5.6.13) for C^+, we have to verify that the perturbation $E = C^+ - C$ does not change the index of C. We shall prove that

$E \in L(W^k_{2,C^+}, W^k_2)$ is the sum of a compact operator and an operator with arbitrarily small norm. Modulo smoothing operators, E can be written

$$E = \sum_{0 \le j < l^+} d_j(t-t_o)^{q^+(j)} D^j P^+ , \; d_j \in C^\infty, \; d_j(t_o)=0 \; (0 \le j < l^+) \qquad (5.6.15)$$

(cf. Remark 2.6.3). Choose $\chi \in C^\infty$ such that $0 \le \chi \le 1$ on Γ, $\chi = 1$ in some neighborhood of t_o and $\chi = 0$ outside a somewhat larger neighborhood U of t_o. Since $\Lambda(1-\chi)E \in L(W^k_{2,C^+}, W^k_2)$ in view of (5.6.15), $(1-\chi)E$ is a compact mapping of W^k_{2,C^+} into W^k_2. Moreover, by (5.6.15) it is therefore sufficient to prove that, for any $d \in C^\infty$ with $d(t_o) = 0$, the multiplication operator $\chi\, dI \in L(W^k_2)$ has arbitrarily small norm modulo compact operators when the diameter of U is small enough. But this is obvious for k = 0 and follows from the fact that

$$\Lambda^k \chi dI - \chi d \Lambda^k \in OP\mathring{C}S^{k-1}$$

if $k \neq 0$. □

<u>5.6.2.</u> We finally give two examples.

<u>Example 5.6.3.</u> Consider the singular integro-differential operator

$$A = \sum_{0 \le j \le l} [\alpha_j(t) + \beta_j(t) S_\Gamma] D^j , \quad \alpha_j, \beta_j \in C^\infty(\Gamma) . \qquad (5.6.16)$$

Setting $b_j^\pm = (\alpha_j \pm \beta_j)/2$, we assume that

$$b_j^\pm = (t-t_o)^j c_j^\pm, \; c_j^\pm \in C^\infty, \; t_o = t(0) \in \Gamma , \; c_l^\pm(t) \neq 0 \; (t \in \Gamma) .$$

Then (5.6.16) can be written $A = B^+P^+ + B^-P^- + K$, where K is a smoothing operator and

$$B^\pm = b_l^\pm D^l + \dots b_1^\pm D + b_o^\pm$$

are Fuchsian differential operators with a singularity at x = 0. The characteristic equations (5.6.6) take the form

$$c_o^\pm(0) + \sum_{1 \le j \le l} c_j^\pm(0)\mu(\mu-1) \dots (\mu-j+1) = 0 , \qquad (5.6.17)$$

and $W^k_{2,A}$ is the weighted Sobolev space

$$W^k_{2,A} = W^k_{2,l} = \{ u \in W^k_2 : (t-t_o)^j D^j u \in W^k_2, j=1,\dots,l \} . \qquad (5.6.18)$$

By Theorem 5.6.1, $A \in L(W^k_{2,A}, W^k_2)$ is a Fredholm operator with index

$$(2\pi)^{-1} \{\arg c_l^-/c_l^+\}_\Gamma - \eta^+(k) + \eta^-(k)$$

if the roots of (5.6.17) satisfy $\mathrm{Re}\,\mu_j^{\pm} \neq -1/2+k$ $(j=1,\dots,l)$. Here $\eta^{\pm}(k)$ denotes the number of roots $\mu_j^{\pm}$ satisfying $\pm\,\mathrm{Re}(\mu_j^{\pm}+1/2-k) < 0$.

Example 5.6.4. Consider the singular integral operator

$$A = \alpha I + \beta S_{\Gamma}\,, \quad \alpha, \beta \in C^{\infty}(\Gamma)\,. \tag{5.6.19}$$

Setting $a^{\pm} = (\alpha \pm \beta)/2$, we suppose that

$$a^{\pm} = (t-t_0)^{q^{\pm}} b^{\pm},\ b^{\pm} \in C^{\infty},\ q^{\pm} \in \mathbb{N}_0,\ t_0 \in \Gamma\,,\ b^{\pm}(t) \neq 0\ (t \in \Gamma\,).$$

Then (5.6.5) reads

$$A = (B^+P^+ + B^-P^-)(\Lambda^{-q^+}P^+ + \Lambda^{-q^-}P^-) + K,\ B^{\pm} = (\mp i)^{q^{\pm}} a^{\pm} D^{q^{\pm}}\,,$$

where K is a smoothing operator. The roots of the characteristic equations (5.6.6) of the Fuchsian differential operators $B^{\pm}$ are $0,1,\dots,q^{\pm}-1$. Moreover,

$$W^k_{2,A} = \{\, u \in \mathcal{D}' : P^{\pm}u \in W^k_{2,q^{\pm}} \} \quad \text{(cf. (5.6.18))}\,.$$

Now it follows from Theorem 5.6.1 that $A \in L(W^k_{2,A}, W^k_2)$ is a Fredholm operator with index

$$(2\pi)^{-1}\{\arg b^-/b^+\}_{\Gamma} \;-\; q^+$$

if $k > \max(q^+, q^-) - 1/2$. By Corollary 5.6.2, (5.6.19) has the same index in C^{∞}.

5.7. Comments and references

5.1. The material in this section is well-known and has been extended to fairly general classes of PDOs on C^{∞} manifolds without boundary; see Hörmander [4], [5], Šubin [1], Beals [1]. Our definition of classical PDOs on a closed curve Γ is due to Agranovič [1] and is of course equivalent to the usual concept, using the Fourier transform for any coordinate neighborhood in Γ.

5.2. Theorem 5.2.1 is a special case of the well-known a priori estimates for elliptic PDOs. The estimate (5.2.3) was first proved by Gårding for partial differential operators and by Kohn and Nirenberg [1] for PDOs. Theorem 5.2.2 is a microlocal version of this inequality in the one-dimensional case. The estimate (5.2.8), together with its microlocal version (5.2.4), is a special case of Melin's inequality proved for PDOs in arbitrary dimensions in Melin [1].

5.3./4. The results of these sections are due to Elschner [9]. Note that, under the assumptions of Corollaries 5.3.10 or 5.4.5, the PDO A is hypoelliptic with loss of one derivative, i.e., for any $k \in \mathbb{R}$, $u \in \mathcal{D}'$ and $Au \in W_2^k$ imply $u \in W_2^{k+l-1}$. The index formula (5.3.19) can be generalized to a class of degenerate PDOs on a closed curve which are hypoelliptic with loss of more than one derivative (see Elschner [10]). Tovmasjan [1] computed the index in C^{∞} of the singular integral operator (5.4.11) in the special case when either $a^+(t) \neq 0$ ($a^-(t) \neq 0$) for all $t \in \Gamma$ or $a^+ = 0$ ($a^- = 0$) and $k^+(t,t) \neq 0$ ($k^-(t,t) \neq 0$) everywhere on Γ. He also considered systems of such operators.

5.5. The method of reducing the oblique derivative problem to a singular integral equation on the boundary Γ is classical. Here we followed the presentation in Michlin [1]. Theorem 5.5.1 contains the well-known index formula in the case when problem (5.5.1) is elliptic, i.e., the Šapiro-Lopatinski condition is satisfied on Γ (see Hörmander [2, Chap. 10], Muschelischwili [1, § 74]). Theorems 5.5.2, 5.5.4 and 5.5.5 are special cases of the results in Elschner [9] on the degenerate oblique derivative problem for general properly elliptic differential operators of second order in the plane. More complete results on the index of this problem may be obtained if the Šapiro-Lopatinski condition degenerates to finite order at isolated points on Γ; see Elschner [6], [7]. In these papers the symbolic calculus of boundary problems for PDOs, due to Boutet de Monvel [1], was used to compute the index by a reduction to a degenerate PDO of first order on the boundary. For a detailed presentation of this calculus and applications to nonelliptic boundary problems, see Rempel and Schulze [1]. With the help of Melin's inequality, an existence and uniqueness result on the degenerate oblique derivative problem for the operator $\Delta + \lambda$, $\lambda < 0$ in dimensions ≥ 2 which is similar to our Theorem 5.5.5 was proved by Taira [1].

5.6. Theorem 5.6.1 may be extended to the case when the principal symbol vanishes to finite order at a finite set of points on $\Gamma \times \{\pm 1\}$;

see Elschner [8]. It can be proved that the degenerate PDO is also a Fredholm operator in C^{∞} and in certain weighted Sobolev spaces when conditions (5.6.8) and (5.6.1o) are not fulfilled. Then the index formula is much more complicated; see Elschner [7], [8] where such a formula was given in terms of the (complete) characteristic factors of certain singular ordinary differential operators. In this case we do not know how to express the index by means of a winding number, using the complete symbol. Using the determining factors (cf. Remark 2.1.7), Kannai [3] proved a criterion for the hypoellipticity of PDOs which satisfy (5.6.1). Index theorems for degenerate PDOs were first obtained in the case of a singular integral operator $\alpha I + \beta S_{\Gamma} + K$, where K is a smoothing operator and the coefficients α, β have various degeneracies on Γ (see Prößdorf [1] and the references therein).

6. A FINITE ELEMENT METHOD FOR PSEUDODIFFERENTIAL EQUATIONS ON A CLOSED CURVE

It is the purpose of this chapter to discuss Galerkin's method using smoothest polynomial splines as trial functions for the approximate solution of pseudodifferential equations on a closed curve. We first recall well-known approximation properties of splines in periodic Sobolev spaces which will then be extended to certain Sobolev spaces with weights. Using Gårding's inequality and standard techniques for finite element methods, we derive the well-known quasioptimal asymptotic error estimates in a range of Sobolev spaces for Galerkin's method with splines when the operator is strongly elliptic. Finally, using Melin's inequality, new results on convergence and error analysis of this method will be given for a class of PDOs which are not strongly elliptic and may degenerate.

6.1. Spline approximation in periodic Sobolev spaces

Let $\Delta = \{x_j\}_{-\infty}^{\infty}$ be a partition of $\mathbb{R}$ by mesh points x_j satisfying $x_0 = -\pi$ and $x_{j+n} = x_j + 2\pi$ for fixed $n \in \mathbb{N}$ and all $j \in \mathbb{Z}$. Let further

$$\overline{h} = \max(x_j - x_{j-1}),\ \underline{h} = \min(x_j - x_{j-1})\ ,\ j=1,\dots,n\ . \tag{6.1.1}$$

By $S_d(\Delta)$ ($d \in \mathbb{N}_0$) we denote the space of all 2π-periodic smoothest splines of degree d subordinate to the partition Δ, i.e., each element of $S_d(\Delta)$ together with its derivatives of order $\leq d-1$ is 2π-periodic and continuous on $\mathbb{R}$, and its restriction to any interval (x_j, x_{j+1}) ($j \in \mathbb{Z}$) is a polynomial of degree $\leq$ d. For any d, $S_d(\Delta)$ is an n-dimensional space and it has a basis $\{B_{j,d}\}_{j=0}^{n-1}$ of so-called B-splines which can be defined in the following way (cf. de Boor [1, Chap. 10]):

Let θ_j be the characteristic function of the interval $[x_j, x_{j+1})$ and define $B_{j,0}$ ($j \in \mathbb{Z}$) by $B_{j,0}(x) = \theta_j(x)$ for $x \in [x_j, x_{j+n})$ and 2π-periodic extension for all $x \in \mathbb{R}$. For $d \geq 1$ and $j \in \mathbb{Z}$, set recurrently

$$B_{j,d}(x) = \frac{x - x_j}{x_{j+d} - x_j} B_{j,d-1}(x) + \frac{x_{j+d+1} - x}{x_{j+d+1} - x_{j+1}} B_{j+1,d-1}(x)$$

for $x \in [x_j, x_{j+n})$ and take the 2π-periodic extension for $x \in \mathbb{R}$.

We have $S_d(\Delta)\subset W_2^s$ if and only if $s< d+1/2$, where W_2^s is the periodic Sobolev space of order s with norm $\|\cdot\|_s$ defined by (5.1.2). The following properties of the spline spaces play an important role in the error analysis of Galerkin methods with splines.

<u>Theorem 6.1.1.</u> (approximation property) If $s\le r\le d+1$, $s< d+1/2$ and $\sigma< s$, then for any $u\in W_2^r$ and any partition Δ there exists $u_\Delta\in S_d(\Delta)$ such that

$$\| u-u_\Delta \|_t \le c(t)\bar{h}^{r-t} \| u\|_r . \tag{6.1.2}$$

for all $t\in[\sigma,s]$, where c(t) denotes a constant independent of u and Δ.

The main step in the proof of Theorem 6.1.1 is to establish

<u>Theorem 6.1.2.</u> If $s\le r\le d+1$ and $s< d+1/2$, then there exists a constant c independent of u and Δ such that

$$\| u-P_{s,\Delta} u\|_s \le c\ \bar{h}^{r-s} \| u\|_r ,\ u\in W_2^r , \tag{6.1.3}$$

where $P_{s,\Delta}$ is the orthogonal projection onto $S_d(\Delta)$ in W_2^s.

<u>Proof.</u> 1. First, we shall verify inequality (6.1.3) for d = 0, r = 1 and $s\in[0,1/2)$. We use the equivalent norms (5.1.5). Let $u\in W_2^1$. We have

$$\varphi(x) := P_{0,\Delta} u = h_j^{-1} \int_{I_j} u(y)dy ,\ x\in I_j\ (j=0,\dots,n-1) ,$$

where $I_j = (x_j,x_{j+1})$ and $h_j = x_{j+1}-x_j$. Let first s = 0. Using the Cauchy-Schwarz inequality, we then obtain

$$\begin{aligned} |u(x)-\varphi(x)|^2 &\le \sum_{0\le j< n} \theta_j(x)|h_j^{-1}\int_{I_j}\{u(x)-u(y)\}\,dy|^2 \\ &\le \sum_{0\le j<n} \theta_j(x)h_j^{-1}\int_{I_j}|u(x)-u(y)|^2dy,\ x\in(-\pi,\pi). \end{aligned} \tag{6.1.4}$$

Furthermore, for any $x,y\in I_j$,

$$\begin{aligned} |u(x)-u(y)|^2 &= |\int_x^y (Du)(t)dt|^2 \le |y-x| \int_{I_j}|Du|^2dx \\ &\le h_j\int_{I_j}|Du|^2dx . \end{aligned} \tag{6.1.5}$$

Integrating this inequality on I_j, we get

$$h_j^{-1}\int_{I_j}|u(x) - u(y)|^2dy \le h_j \int_{I_j}|Du|^2dx .$$

Integrating the last estimate on I_j and using (6.1.4), we find that

$$\|u-\varphi\|_0^2 \leq \sum_{0\leq j<n} h_j^2 \int_{I_j} |Du|^2 dx \leq \bar{h}^2 \|Du\|_0^2 \tag{6.1.6}$$

which implies (6.1.3) for $d = s = 0$ and $r = 1$.

Let now $s \in (0,1/2)$. Then in light of (6.1.6), inequality (6.1.3) is a consequence of the following estimate:

$$\begin{aligned} J &:= \int_{-\pi}^{\pi}\int_{-\pi}^{\pi} |u(x)-\varphi(x)-u(y)+\varphi(y)|^2 |x-y|^{-1-2s} dxdy \\ &\leq c\, \bar{h}^{2-2s} \|Du\|_0^2 . \end{aligned} \tag{6.1.7}$$

Henceforth the letter c will denote various positive constants independent of u and Δ . We have

$$J = \sum_{0\leq j,k<n} J_{jk} ,$$

where J_{jk} denotes the corresponding integral on the rectangle $I_{jk} = I_j \times I_k$. Using (6.1.5), we obtain

$$\begin{aligned} J_{jj} &= \int_{I_{jj}} |u(x)-u(y)|^2 |x-y|^{-1-2s} dxdy \\ &\leq \int_{I_j} |Du|^2 dx \int_{I_{jj}} |x-y|^{-2s} dxdy \leq c\, \bar{h}^{2-2s} \int_{I_j} |Du|^2 dx, \quad j=0,\dots,n-1. \end{aligned} \tag{6.1.8}$$

Applying (6.1.4) and (6.1.5), for $j\neq k$ we get

$$\begin{aligned} J_{jk} &\leq 2 \Big\{ \int_{I_{jk}} |u(x)-\varphi(x)|^2 |x-y|^{-1-2s} dxdy \\ &+ \int_{I_{jk}} |u(y)-\varphi(y)|^2 |x-y|^{-1-2s} dxdy \Big\} \leq 2\bar{h} \Big\{ \int_{I_j} |Du|^2 dx \\ &+ \int_{I_k} |Du|^2 dx \Big\} \int_{I_{jk}} |x-y|^{-1-2s} dxdy . \end{aligned}$$

This implies

$$\sum_{j\neq k} J_{jk} \leq 8\bar{h} \sum_{0\leq j<n} \Big(\sum_{j<k<n} c_{jk} \Big) \int_{I_j} |Du|^2 dx , \tag{6.1.9}$$

$$\begin{aligned} c_{jk} &= \int_{I_k}\int_{I_j} (y-x)^{-1-2s} dxdy = (2s(1-2s))^{-1} \big\{ (x_{k+1}-x_{j+1})^{1-2s} \\ &+ (x_k-x_j)^{1-2s} - (x_{k+1}-x_j)^{1-2s} - (x_k-x_{j+1})^{1-2s} \big\} . \end{aligned}$$

Finally, combining the estimate

$$\sum_{j<k<n} c_{jk} = (2s(1-2s))^{-1} \{ (x_{j+1}-x_j)^{1-2s} + (x_n-x_{j+1})^{1-2s}$$
$$- (x_n-x_j)^{1-2s} \} \le (2s(1-2s))^{-1}(x_{j+1}-x_j)^{1-2s}$$

with (6.1.8) and (6.1.9), one obtains (6.1.7) Q.E.D.

2. Next, we prove (6.1.3) for $r = d+1$, $s \in [d,d+1/2)$. Introduce the spaces

$$\mathring{W}_2^s = \{ u \in W_2^s : \int_{-\pi}^{\pi} u dx = 0 \}, \; s \ge 0, \; \mathring{S}_d(\Delta) = S_d(\Delta) \cap \mathring{W}_2^0 .$$

Then $W_2^s = \mathring{W}_2^s \dotplus \mathbb{C}$, $S_d(\Delta) = \mathring{S}_d(\Delta) \dotplus \mathbb{C}$, and it is easy to check that the operator D^d is an isomorphism of $\mathring{W}_2^{s+d}$ and $\mathring{S}_d(\Delta)$ onto $\mathring{W}_2^s$ and $\mathring{S}_o(\Delta)$, respectively. Therefore it is sufficient to verify that, for any $u \in \mathring{W}_2^{d+1}$, there exists $\psi \in \mathring{S}_o(\Delta)$ such that

$$\| D^d u - \psi \|_{s-d} \le c \, \bar{h}^{d+1-s} \| D^d u \|_1 ,$$

which follows immediately from the first part of the proof.

3. We now prove that (6.1.3) holds for all integers s and r satisfying $s \le r \le d+1$ and $s \le d$. This follows from 2. when $r = d+1$ and $s = d$ and is trivial when $r = s$. Suppose the assertion is true for all $s \ge j$ and $r > s$, where j is some integer $\le d$. Using (5.6.1) and the assumption, we obtain

$$\| u-P_{j-1,\Delta} u \|_{j-1} \le \| u-P_{j,\Delta} u \|_{j-1} = \sup |(\psi, u-P_{j,\Delta} u)_j|$$
$$\le \sup \| \psi - P_{j,\Delta} \psi \|_j \, \| u-P_{j,\Delta} u \|_j \le c \, \bar{h}^{r+1-j} \| u \|_r , \qquad (6.1.10)$$

where the supremum is taken over all $\psi \in W_2^{j+1}$ with norm ≤ 1. Thus, by induction on $j = d, d-1, \ldots$, the assertion is proved.

4. We complete the proof of Theorem 6.1.2 by interpolation. In view of (6.1.10), $I-P_{j,\Delta}$ $(j=r-1,r-2,\ldots;r=d+1,d,\ldots)$ is a continuous map of W_2^r into W_2^{j-1} and of W_2^r into W_2^j with norm $\le c \, \bar{h}^{r+1-j}$ and $\le c \, \bar{h}^{r-j}$, respectively. Therefore (5.1.8) implies (6.1.3) for $s \in (j-1,j)$. Moreover, interpolation for the operators

$$I-P_{s,\Delta} : W_2^s \to W_2^{d+1} \text{ (resp. } W_2^r)$$

yields (6.1.3) for all $s \le d$ and $s \le r \le d+1$. Finally, by interpolation applied to the maps

$$I-P_{s,\Delta'} : W_2^s \to W_2^{d+1} \quad \text{(resp. } W_2^s) ,$$

we obtain Theorem 6.1.2 in the case $d < s < d+1/2$, $s \le r \le d+1$. □

<u>Proof of Theorem 6.1.1.</u> Let $u \in W_2^s$ and

$$E(u) = \inf_{\varphi \in S_d(\Delta)} \{ \|u-\varphi\|_\sigma + \bar{h}^{s-\sigma} \|u-\varphi\|_s \}, \beta = \sup_{\|u\|_s \le 1} E(u) .$$

From the definition of β and Theorem 6.1.2 it follows that

$$\begin{aligned} E(v) &= \inf_{\psi \in S_d(\Delta)} E(v-\psi) \le \beta \inf_{\psi \in S_d(\Delta)} \|v-\psi\|_s \\ &\le c\, \bar{h}^{r-s} \beta \|v\|_r , \quad v \in W_2^r . \end{aligned} \tag{6.1.11}$$

Hence, for any $v \in W_2^r$,

$$E(u) \le E(u-v)+E(v) \le \|u-v\|_\sigma + \bar{h}^{s-\sigma}\|u-v\|_s + c\, \bar{h}^{r-s}\beta \|v\|_r, \tag{6.1.12}$$

where we have used (6.1.11), the triangle inequality and the fact that $0 \in S_d(\Delta)$. Furthermore, for any $u \in W_2^s$ and $\varepsilon > 0$, there exists $v \in W_2^r$ such that

$$\|u-v\|_s \le \|u\|_s, \ \|u-v\|_\sigma \le \varepsilon^{s-\sigma}\|u\|_s, \ \|v\|_r \le \varepsilon^{s-r} \|u\|_s . \tag{6.1.13}$$

Indeed, setting

$$v = (2\pi)^{-1/2} \sum_{|m| \le \bar{m}} \hat{u}_m e^{im x} ,$$

where $\bar{m}$ is the greatest natural number $\le 1/\varepsilon$ and $\hat{u}_m$ are the Fourier coefficients of u, one obtains (6.1.13) immediately from the definition (5.1.2) of the norm. With the choice of v according to (6.1.13), (6.1.12) gives

$$E(u) \le \{ \bar{h}^{s-\sigma} + \varepsilon^{s-\sigma} + c(\bar{h}/\varepsilon)^{r-s} \beta \} \|u\|_s .$$

Choosing ε such that $c(\bar{h}/\varepsilon)^{r-s} \le 1/2$ and taking the supremum over all $\|u\|_s \le 1$, we get $\beta \le c\, \bar{h}^{s-\sigma}$. Together with (6.1.11), this implies

$$E(u) \le c\, \bar{h}^{r-\sigma}\|u\|_r , \quad u \in W_2^r .$$

Hence, for any $u \in W_2^r$ and any partition Δ, there exists $u_\Delta \in S_d(\Delta)$ such that estimate (6.1.2) holds for $t = \sigma$ and $t = s$. Finally, using the inequality

$$\|u\|_t \le \|u\|_\sigma^{1-\theta} \|u\|_s^\theta, \ u \in W_2^s, \ t = (1-\theta)\sigma + \theta s, \ 0 < \theta < 1 \tag{6.1.14}$$

which is an immediate consequence of (5.1.2) and Hölder's inequality, we obtain (6.1.2) for $t \in (\sigma, s)$. □

A mesh Δ is said to be γ-quasiuniform ($0<\gamma\leq 1$) if $\gamma\bar{h}\leq \underline{h}$, and the set of all γ-quasiuniform meshes is denoted by $\mathcal{D}_\gamma$.

<u>Theorem 6.1.3.</u> (inverse property) For any real numbers t, s and γ satisfying $t\leq s<d+1/2$ and $0<\gamma\leq 1$, there exists a constant c such that

$$\|\varphi\|_s \leq c\,\bar{h}^{t-s}\,\|\varphi\|_t \qquad (6.1.15)$$

for all $\varphi\in S_d(\Delta)$ and $\Delta\in\mathcal{D}_\gamma$.

<u>Proof.</u> We first verify (6.1.15) for $d<t\leq s<d+1/2$. We use the equivalent norms (5.1.5). Since $D^d\varphi\in S_0(\Delta)$ for $\varphi\in S_d(\Delta)$, it remains to prove the estimate

$$J_s(\varphi)\leq c\,\bar{h}^{2(t-s)}J_t(\varphi)\ ,\quad \varphi\in S_0(\Delta),\ \Delta\in\mathcal{D}_\gamma\ , \qquad (6.1.16)$$

where $0<t\leq s<1/2$ and

$$J_s(\varphi)=\int_{-\pi}^{\pi}\int_{-\pi}^{\pi}|\varphi(x)-\varphi(y)|^2|x-y|^{-1-2s}dx\,dy\ .$$

Let $\varphi=c_0\theta_0+\ldots+c_{n-1}\theta_{n-1}, c_j\in\mathbb{C}$. Recall that θ_j is the characteristic function of the interval $I_j=[x_j,x_{j+1})$. We have

$$J_s(\varphi)=\sum_{0\leq j,k<n}J_{jk,s}\ ,\quad J_{jk,s}=\int_{I_j}\int_{I_k}|c_j-c_k|^2|x-y|^{-1-2s}dxdy, \qquad (6.1.17)$$

and for $|j-k|\geq 2$,

$$J_{jk,s}=\int_{I_j}\int_{I_k}\frac{|c_j-c_k|^2}{|x-y|^{1+2t}}\ \frac{dxdy}{|x-y|^{2(s-t)}}\leq \underline{h}^{2(t-s)}J_{jk,t}\ , \qquad (6.1.18)$$

while for $k=j+1$,

$$J_{jj+1,s}=|c_{j+1}-c_j|^2(2s(1-2s))^{-1}\{h_j^{1-2s}+h_{j+1}^{1-2s}-(h_j+h_{j+1})^{1-2s}\}. \qquad (6.1.19)$$

Therefore

$$J_{jj+1,s}/J_{jj+1,t}\leq c\,h_j^{2(t-s)}f_s(\bar{x})/f_t(\bar{x})\ ,$$

where $\bar{x}=h_{j+1}/h_j$ and $f_s(x)=1+x^{1-2s}-(1+x)^{1-2s}$. For any $s\in(0,1/2)$, f_s is increasing on $(0,\infty)$. Moreover, since $\Delta\in\mathcal{D}_\gamma$, we have $\bar{x}\in[\gamma,1/\gamma]$. Thus we get

$$J_{jj+1,s}/J_{jj+1,t}\leq c\{f_s(1/\gamma)/f_t(\gamma)\}\,\underline{h}^{2(t-s)}$$

which implies an estimate of the form (6.1.18) again. By virtue of

(6.1.17) and $\Delta \in \mathcal{D}_\gamma$, the proof of inequality (6.1.16) is complete.

2. We now extend (6.1.15) to the general case. For $d < t < s < d+1/2$, we obtain

$$\|\varphi\|_t^2 \le \|\varphi\|_s \, \|\varphi\|_{2t-s} \le c\, \bar{h}^{t-s} \|\varphi\|_t \, \|\varphi\|_{2t-s}$$

by (6.1.14) and the first step of the proof. Using the estimate $\|\varphi\|_s \le c\, \bar{h}^{t-s} \|\varphi\|_t$, this gives $\|\varphi\|_s \le c\, \bar{h}^{2t-2s} \|\varphi\|_{2t-s}$. By induction,

$$\|\varphi\|_s \le c\, \bar{h}^{k(t-s)} \|\varphi\|_{t+k(t-s)}, \quad k=1,2,\ldots,$$

hence (6.1.15) for $s \in (d,d+1/2)$ and any $t \le s$. If $s \le d$, then we choose some $r \in (d,d+1/2)$ and using (6.1.14) again, we obtain

$$\|\varphi\|_s \le \|\varphi\|_t^{(r-s)/(r-t)} \|\varphi\|_r^{(s-t)/(r-t)}$$

$$\le c\|\varphi\|_t^{(r-s)/(r-t)} \bar{h}^{(t-r)(s-t)/(r-t)} \|\varphi\|_t^{(s-t)/(r-t)} \le c\, \bar{h}^{t-s}\|\varphi\|_t. \square$$

6.2. Spline approximation in weighted Sobolev spaces

Let $a \in C^\infty$ such that $\mathrm{Re}\, a(x) \ge 0$ $(x \in \mathbb{R})$. We introduce the weighted Sobolev spaces

$$W^k_{2,a} = \{ u \in W^k_2 : au \in W^{k+1}_2 \}, \; k \in \mathbb{R}$$

endowed with the canonical norm $\|u\|_{k,a} = \|u\|_k + \|au\|_{k+1}$. Let further $A^\mu = a\Lambda + \mu$, where $\mu \in \mathbb{C}$ and Λ is the PDO defined in 5.1.1. Since $\|u\|_k + \|a\Lambda u\|_k$ is obviously equivalent to the norm $\|u\|_{k,a}$, $W^k_{2,a}$ coincides with the space W^k_{2,A^μ} defined in 5.3.1.

Lemma 6.2.1. For all sufficiently large $\mu > 0$, A^μ is an isomorphism of $W^k_{2,a}$ onto W^k_2. Moreover, for those μ, $|||u|||_{k,a} = \|A^\mu u\|_k$ is equivalent to the norm $\|u\|_{k,a}$ and $W^k_{2,a}$ is a Hilbert space with the scalar product

$$(u,v)_{k,a} = (A^\mu u, A^\mu v)_k . \tag{6.2.1}$$

Proof. The PDO A^μ has the symbol $a|\xi| + \mu$, hence the corresponding function a^k_o defined in (5.3.2) takes the form

$$a^k_o = \mu + (k-1/2)(i\xi / |\xi|)^{-1} Da .$$

Therefore, Re $a_0^k(x,\pm 1) > 0$ for all x when μ is sufficiently large, and the lemma follows from Theorem 5.3.3. □

In view of (5.1.6) and (6.2.1), for any $j,k \in \mathbb{R}$ and $u \in W_{2,a}^{j-k}$,

$$|||u|||_{j-k,a} = \sup_{v \in W_{2,a}^{j+k}, v \neq 0} |(u,v)_{j,a}| / |||v|||_{j,a} \tag{6.2.2}$$

with suitable $\mu > 0$.

The aim of this section is to extend Theorem 6.1.1 to weighted Sobolev spaces.

<u>Theorem 6.2.2.</u> If $s \leq r \leq d$, $s < d-1/2$ and $\sigma < s$, then for any $u \in W_{2,a}^r$ and any mesh Δ there exists $u_\Delta \in S_d(\Delta)$ such that

$$\| u-u_\Delta \|_{t,a} \leq c(t)\bar{h}^{r-t} \| u \|_{r,a}$$

for all $t \in [\sigma, s]$, where the constant c(t) is independent of u and Δ.

We first prove an analogue of Theorem 6.1.2. Let $Q_{s,\Delta}$ be the orthogonal projection of $W_{2,a}^s$ onto $S_d(\Delta)$ with respect to the scalar product (6.2.1) for some suitable $\mu > 0$. Henceforth by c we denote various constants independent of u and Δ.

<u>Theorem 6.2.3.</u> For $s \leq r \leq d$ and $s < d-1/2$,

$$\| u-Q_{s,\Delta} u \|_{s,a} \leq c\, \bar{h}^{r-s} \| u \|_{r,a}, \quad u \in W_{2,a}^r. \tag{6.2.3}$$

<u>Proof.</u> 1. First, we verify (6.2.3) for d = 0, r = 0 and $s \in [-1,-1/2)$. As in the proof of Theorem 6.1.2, we set $\varphi = P_{0,\Delta} u$ for $u \in W_{2,a}^0 \subset L_2$. Let s = -1. We shall show the inequality

$$\| u-\varphi \|_{-1} + \| au-a\varphi \|_0 \leq c\, \bar{h} \| u \|_{0,a} \tag{6.2.4}$$

which implies (6.2.3) in this case. It follows from the proof of Theorem 6.1.2 that

$$\| u-\varphi \|_{-1} \leq c\, \bar{h} \| u \|_0, \quad \| au-P_{0,\Delta} au \|_0 \leq c\, \bar{h} \| au \|_1 \leq c\, \bar{h} \| u \|_{0,a}.$$

We now deduce the estimate

$$\| a\varphi - P_{0,\Delta} au \|_0 \leq c\, \bar{h} \| u \|_0 \tag{6.2.5}$$

which completes the proof of (6.2.4). We have

$$\Psi(x) := a\varphi - P_{0,\Delta} au = h_j^{-1} \int_{I_j} \{a(x)-a(y)\}\, u(y)dy,$$

$x \in I_j$, $j=0,\dots,n-1$,

where $I_j = (x_j, x_{j+1})$ and $h_j = x_{j+1} - x_j$. By the Cauchy-Schwarz inequality one obtains

$$|\psi(x)|^2 \leq h_j^{-1} \int_{I_j} |a(x)-a(y)|^2 |u(y)|^2 dy \leq c\, h_j \int_{I_j} |u|^2 dy, \quad x \in I_j \tag{6.2.6}$$

since

$$|a(x)-a(y)| \leq c\,|x-y| \leq c\, h_j , \quad x,y \in I_j . \tag{6.2.7}$$

Integrating inequality (6.2.6) on I_j and summing up with respect to j, we get (6.2.5).

Let now $s \in (-1,-1/2)$. We shall verify the inequality

$$\| u-\varphi \|_s + \| au-a\varphi \|_{s+1} \leq c\, \bar{h}^{-s} \| u \|_{0,a} \tag{6.2.8}$$

which proves (6.2.3) in this case. It follows from the proof of Theorem 6.1.2 that

$$\| u-\varphi \|_s \leq c\, \bar{h}^{-s} \| u \|_0 , \quad \| au-P_{0,\Delta} au \|_{s+1} \leq c\, \bar{h}^{-s} \| au \|_1 \leq c\, \bar{h}^{-s} \| u \|_{0,a}.$$

To prove (6.2.8), it remains to verify the estimate

$$\| \psi \|_{s+1} \leq c\, \bar{h}^{-s} \| u \|_0 .$$

By virtue of (6.2.5) it suffices to show that

$$J := \int_{-\pi}^{\pi} \int_{-\pi}^{\pi} |\psi(x)-\psi(y)|^2 |x-y|^{-3-2s} dx dy \leq c\, \bar{h}^{-2s} \| u \|_0^2 . \tag{6.2.9}$$

We have

$$J = \sum_{0 \leq j,k < n} J_{jk} ,$$

where J_{jk} denotes the corresponding integral on the rectangle $I_{jk} = I_j \times I_k$. By (6.2.7),

$$\begin{aligned} J_{jj} &\leq |h_j^{-1} \int_{I_j} u\, dx|^2 \int_{I_{jj}} |a(x)-a(y)|^2 |x-y|^{-3-2s} dx dy \\ &\leq c\, h_j^{-1} \int_{I_j} |u|^2 dx \int_{I_{jj}} |x-y|^{-1-2s} dx dy \leq c\, \bar{h}^{-2s} \int_{I_j} |u|^2 dx . \end{aligned} \tag{6.2.10}$$

Applying (6.2.6), for $j \neq k$ we claim

$$J_{jk} \leq 2 \left\{ \int_{I_{jk}} (|\psi(x)|^2 + |\psi(y)|^2) |x-y|^{-3-2s} dx dy \right\}$$

$$\leq 2\bar{h}\left\{ \int_{I_j} |u|^2 dx + \int_{I_k} |u|^2 dy \right\} \int_{I_{jk}} |x-y|^{-3-2s} dx dy \, . \qquad (6.2.11)$$

As in the first step of the proof of Theorem 6.1.2, (6.2.1o) and (6.2.11) imply (6.2.9).

2. Next, we consider the case $r = d$ and $s \in [d-1, d-1/2)$. Introduce the subspace $\mathring{W}^s_{2,a} = \{ u \in W^s_{2,a} : \langle u,1 \rangle = 0 \}$ of $W^s_{2,a}$, where $\langle \cdot,\cdot \rangle$ denotes the duality between C^∞ and $\mathcal{D}'$. D^d is an isomorphism of $\mathring{W}^s_{2,a}$ onto $\mathring{W}^{s-d}_{2,a}$ for any $s \in \mathbb{R}$. As in the second step of the proof of Theorem 6.1.2, one can now reduce the assertion to the case $r = 0$, $s \in [-1,-1/2)$.

3. Using (6.2.2) and replacing the projections $P_{s,\Delta}$ by $Q_{s,\Delta}$, the method in the third part of the proof of Theorem 6.1.2 yields (6.2.3) for all integers s and r satisfying $s \leq d-1$ and $s \leq r \leq d$.

4. Using the equivalent norms $|||\cdot|||_{s,a}$, we observe that the interpolation inequality (5.1.8) extends to the weighted Sobolev spaces. As in the fourth step of the proof of Theorem 6.1.2, we obtain (6.2.3) for all s and r by interpolation. □

Proof of Theorem 6.2.2. By Lemma 6.2.1, we may choose $\mu > 0$ so large that A^μ is an isomorphism of $W^k_{2,a}$ onto W^k_2 for all $k \in [\sigma, r]$. With $u = A^\mu \tilde{u}$ and the choice of v according to (6.1.13), we observe that, for any $\tilde{u} \in W^s_{2,a}$ and $\varepsilon > 0$, there exists $w = (A^\mu)^{-1} v \in W^r_{2,a}$ satisfying

$$|||\tilde{u}-w|||_{s,a} \leq |||\tilde{u}|||_{s,a},\ |||\tilde{u}-w|||_{\sigma,a} \leq \varepsilon^{s-\sigma} |||\tilde{u}|||_{s,a},\ |||w|||_{r,a} \leq$$
$$\leq \varepsilon^{s-r} |||\tilde{u}|||_{s,a} \, .$$

Moreover, we see that inequality (6.1.14) extends to the norms $|||\cdot|||_{s,a}$. Thus, as in the proof of Theorem 6.1.1, one can deduce Theorem 6.2.2 from Theorem 6.2.3. □

6.3. The Galerkin method with splines for strongly elliptic equations

Let Γ be a simple closed C^∞ curve in the plane which has the equation $t = t(x)$, $x \in [-\pi, \pi]$. We identify functions on Γ with 2π-periodic functions on the real axis (cf. (5.1.1)). Consider a classical PDO A of order $l \in \mathbb{R}$ on Γ with symbol

$$\sigma_A(x,\xi) = a_l(x,\xi) + a_{l-1}(x,\xi) + \dots \quad . \tag{6.3.1}$$

We shall assume throughout this section that A is strongly elliptic:

$$\operatorname{Re} a_l(x,\xi) > 0 \quad \text{on} \quad \mathbb{R} \times \{\pm 1\} . \tag{6.3.2}$$

It follows from Theorem 5.1.14 and Corollary 5.1.15 that then $A : W_2^k \to W_2^{k-l}$ is a continuous Fredholm operator with index 0 for any $k \in \mathbb{R}$, and its invertibility is equivalent to the implication

$$u \in C^\infty , \; Au = 0 \Rightarrow u = 0 . \tag{6.3.3}$$

Consider the equation

$$Au = f, \; f \in W_2^{-l/2} . \tag{6.3.4}$$

The Galerkin method with splines for the approximate solution of equation (6.3.4) seeks an element $u_\Delta \in S_d(\Delta)$ satisfying the Galerkin equations

$$(Au_\Delta , v_\Delta)_0 = (f, v_\Delta)_0 \text{ for all } v_\Delta \in S_d(\Delta) . \tag{6.3.5}$$

Here $(\cdot,\cdot)_0$ denotes the scalar product in $L_2(-\pi,\pi)$ (resp. its extension to an antiduality between $W_2^{l/2}$ and $W_2^{-l/2}$). Using the basis of B-splines for $S_d(\Delta)$ defined in Sec. 6.1, we see that (6.3.5) is equivalent to a linear system of n equations in n unknowns, where n denotes the number of mesh points in $[-\pi,\pi)$. The matrix of this system is sparse if A is a differential operator.

The following theorem establishes the convergence of the Galerkin method in the "natural energy space" $W_2^{l/2}$ for A. Let $\bar{h}$ be the mesh size of Δ; see (6.1.1).

<u>Theorem 6.3.1.</u> Assume (6.3.2), (6.3.3) and $l/2 < d+1/2$. If $\bar{h}$ is sufficiently small, then equations (6.3.5) are uniquely solvable for any $f \in W_2^{-l/2}$, and the approximate solutions u_Δ converge in $W_2^{l/2}$ to the exact solution u of (6.3.4) with the error bound

$$\| u-u_\Delta \|_{l/2} \leq c \min_{v \in S_d(\Delta)} \| u-v \|_{l/2} \quad (\bar{h} \to 0) . \tag{6.3.6}$$

<u>Proof.</u> Let P_Δ be the orthogonal projection of $W_2^{l/2}$ onto the spline space $S_d(\Delta)$, and $P_\Delta^* \in L(W_2^{-l/2})$ the adjoint of P_Δ with respect to the L_2 inner product. Then equations (6.3.5) can be written as a projection

method with respect to the projections P_Δ and P_Δ^*:

$$A_\Delta u_\Delta := P_\Delta^* A P_\Delta u_\Delta = P_\Delta^* f . \tag{6.3.7}$$

We first show the stability of the method, i.e. the estimate

$$\| A_\Delta u_\Delta \|_{-1/2} \geq c \| u_\Delta \|_{1/2} \tag{6.3.8}$$

for all $u_\Delta \in S_d(\Delta)$ and sufficiently small $\bar{h}$. Otherwise there would exist a sequence of meshes $\{\Delta_j\}$ and a sequence $\{u_j\}$ of elements $u_j \in S_d(\Delta_j)$ such that the mesh size of Δ_j tends to zero, $\|u_j\|_{1/2} = 1$ and $\| A_{\Delta_j} u_j \|_{-1/2} \to 0$ as $j \to \infty$. Since the unit ball in $W_2^{1/2}$ is weakly compact, we can assume that u_j converges weakly in $W_2^{1/2}$ to some element u. We show that u = 0. By Theorem 6.1.2 and the Banach-Steinhaus theorem, P_{Δ_j} converges strongly to the identity operator in $W_2^{1/2}$. Consequently,

$$(A_{\Delta_j} u_j, v)_0 = (u_j, A^* P_{\Delta_j} v)_0 \to (u, A^* v)_0 = (Au, v)_0 \quad (j \to \infty)$$

for any $v \in W_2^{1/2}$, where $A^* \in L(W_2^{1/2}, W_2^{-1/2})$ is the formal adjoint of the PDO A. Hence, in light of (5.1.6), $A_{\Delta_j} u_j$ converges weakly to Au in $W_2^{-1/2}$. On the other hand, $A_{\Delta_j} u_j \to 0$ in $W_2^{-1/2}$ by assumption. Thus the invertibility of $A \in L(W_2^{1/2}, W_2^{-1/2})$ implies u = 0.

Furthermore, applying Gårding's inequality (Corollary 5.2.3), we obtain

$$\begin{aligned} 0 < c = c \|u_j\|_{1/2}^2 &\leq \|u_j\|_{(1-1)/2}^2 + \mathrm{Re}(A_{\Delta_j} u_j, u_j)_0 \\ &\leq \|u_j\|_{(1-1)/2}^2 + \| A_{\Delta_j} u_j \|_{-1/2} \end{aligned} \tag{6.3.9}$$

for all j. Since the embedding $W_2^{1/2} \subset W_2^{(1-1)/2}$ is compact, the right-hand side of (6.3.9) converges to 0 as $j \to \infty$, hence a contradiction which proves (6.3.8).

It remains to prove (6.3.6) since we have already observed that P_Δ converges strongly to the identity in $W_2^{1/2}$ as $\bar{h} \to 0$. By (5.1.6), P_Δ^* converges strongly to the identity in $W_2^{-1/2}$, too. Therefore the norm of the operators $P_\Delta^* A(I-P_\Delta) \in L(W_2^{1/2}, W_2^{-1/2})$ is uniformly bounded by the Banach-Steinhaus theorem. Since

$$A_\Delta (u_\Delta - P_\Delta u) = P_\Delta^* A (I - P_\Delta) u$$

in view of (6.3.4) and (6.3.7), it follows from (6.3.8) that

$$\| u_\Delta - P_\Delta u \|_{1/2} = \| A_\Delta^{-1} P_\Delta^* A(I-P_\Delta) u \|_{1/2} \le c \, \| u - P_\Delta u \|_{1/2}$$

for all sufficiently small $\bar{h}$. Combining the last estimate with the triangle inequality

$$\| u - u_\Delta \|_{1/2} \le \| u_\Delta - P_\Delta u \|_{1/2} + \| u - P_\Delta u \|_{1/2} ,$$

we finally obtain (6.3.6). □

Using Theorem 6.1.1, we further obtain asymptotic error bounds in a range of Sobolev spaces.

<u>Theorem 6.3.2.</u> Assume (6.3.2), (6.3.3), $s \le r \le d+1$, $s < d+1/2$, $1/2 < d+1/2$, $1/2 \le r$, $f \in W_2^{r-1}$ and, if $s > 1/2$, $\Delta \in \mathcal{D}_\gamma$. Then for any sufficiently small $\bar{h}$,

$$\| u - u_\Delta \|_s \le c \, \| u \|_r \begin{cases} \bar{h}^{r-s} & \text{if } d \ge l-1-s , \\ \bar{h}^{d+r-l+1} & \text{if } d \le l-1-s , \end{cases} \tag{6.3.10}$$

where u is the solution of (6.3.4) and u_Δ the solution of (6.3.5).

<u>Proof.</u> Since $f \in W_2^{r-1}$, we have $u \in W_2^r$ by Corollary 5.1.15. We now consider three cases.

1. Let $s = 1/2 \le r$. Then (6.3.10) follows from (6.3.6) and Theorem 6.1.2.

2. Suppose $1/2 < s \le r$. Using Theorem 6.1.3, for arbitrary $\chi \in S_d(\Delta)$ one obtains

$$\| u - u_\Delta \|_s \le \| u - \chi \|_s + \| u_\Delta - \chi \|_s \le \| u - \chi \|_s + c \, \bar{h}^{1/2-s} \| u_\Delta - \chi \|_s . \tag{6.3.11}$$

Since the norm of the operators $A_\Delta^{-1} P_\Delta^* A \in L(W_2^{1/2})$ is uniformly bounded for all sufficiently small $\bar{h}$ (see the proof of Theorem 6.3.1), we claim that

$$\| u_\Delta - \chi \|_{1/2} = \| A_\Delta^{-1} P_\Delta^* A(u - \chi) \|_{1/2} \le c \, \| u - \chi \|_{1/2} .$$

Together with (6.3.11), this implies

$$\| u - u_\Delta \|_s \le \| u - \chi \|_s + c \, \bar{h}^{1/2-s} \| u - \chi \|_{1/2} , \quad \chi \in S_d(\Delta) . \tag{6.3.12}$$

By Theorem 6.1.1, there exists $\chi \in S_d(\Delta)$ such that the right-hand side of (6.3.12) can be estimated by $c \, \bar{h}^{r-s}$ which proves (6.3.10).

3. Let $s<1/2$. Since $A \in L(W_2^s, W_2^{s-1})$ is invertible in view of (6.3.2) and (6.3.3), $A^* \in L(W_2^{1-s}, W_2^{-s})$ is invertible, too. Because of (5.1.6) we can write

$$\| u-u_\Delta \|_s = \sup_{g \in W_2^{-s}, g \neq 0} |(u-u_\Delta, g)_o| / \| g \|_{-s} . \tag{6.3.13}$$

Let $w \in W_2^{1-s}$ be the unique solution of the equation $A^* w = g$. Since

$$(u-u_\Delta, g)_o = (u-u_\Delta, A^* w)_o = (u-u_\Delta, A^*(w-v))_o, \quad v \in S_d(\Delta) ,$$

we obtain the inequality

$$|(u-u_\Delta, g)_o| \leq c \min_{v \in S_d(\Delta)} \| w-v \|_{1/2} \| u-u_\Delta \|_{1/2} . \tag{6.3.14}$$

Furthermore, Theorem 6.1.2 implies

$$\min_{v \in S_d(\Delta)} \| w-v \|_{1/2} \leq c \begin{cases} \bar{h}^{1/2-s} & \| w \|_{1-s} \text{ if } d \geq l-1-s, \\ \bar{h}^{d-1/2+1} & \| w \|_{d+1} \text{ if } d \leq l-1-s. \end{cases} \tag{6.3.15}$$

Since $\| w \|_{1-s} \leq c \| g \|_{-s}$ and $\| u-u_\Delta \|_{1/2} \leq c\, \bar{h}^{r-1/2} \| u \|_r$ (see the first step of the proof), we obtain (6.3.1o) from (6.3.13) to (6.3.15).□

<u>Remark 6.3.3.</u> It follows from Theorem 6.3.2 that the maximal rate of convergence achieved by the Galerkin method is $O(\bar{h}^{2d+2-l})$ in W_2^{l-d-1}.

<u>Example 6.3.4.</u> Consider the PDO $A = V+K$, where V and K are defined as in (5.5.6). A is a strongly elliptic PDO with principal symbol $|\xi|^{-1}$. Assume (6.3.3). Then we obtain from Theorems 6.3.1 and 6.3.2 that the Galerkin method (6.3.5) converges in $W_2^{-1/2}$ for any right-hand side $f \in W_2^{1/2}$ and, for $s \leq r \leq d+1$, $s < d+1/2$, $-1/2 \leq r$, $f \in W_2^{r+1}$ and $\Delta \in \mathcal{D}_\gamma$ if $s > -1/2$,

$$\| u-u_\Delta \|_s \leq c \| u \|_r \begin{cases} \bar{h}^{r-s} & \text{when } d \geq -2-s , \\ \bar{h}^{d+r+2} & \text{when } d \leq -2-s . \end{cases}$$

<u>6.4. The Galerkin method with splines for degenerate equations</u>

<u>6.4.1.</u> Consider a classical PDO A on Γ with symbol (6.3.1) and subprincipal symbol $a'_{l-1} = a_{l-1} - (2i)^{-1} D_x D_\xi a_l$. In this section we extend Theorems 6.3.1 and 6.3.2 to the case when the strong ellipticity is replaced by the following two conditions:

$$\begin{aligned} &\text{Re } a_l \geq 0 \quad \text{on} \quad \mathbb{R} \times \{\pm 1\} \,, \\ &\text{Re } a'_{l-1} > 0 \text{ on } \{(x,\pm 1) : x \in \mathbb{R},\ \text{Re } a_l(x,\pm 1) = 0\} \,. \end{aligned} \tag{6.4.1}$$

Let

$$W^k_{2,A} = \{u \in W_2^{k+l-1} : a^{\pm} D P^{\pm} u \in W_2^{k+l-1}\},\quad a^{\pm}(x) = a_l(x,\pm 1)$$

be the weighted Sobolev space defined in 5.3.1. Then we have

$$W^k_{2,A} = \{u \in W_2^{k+l-1} : P^{\pm} u \in W^{k+l-1}_{2,a}\};$$

$W^k_{2,A}$ is clearly a Hilbert space with scalar product

$$(P^+u, P^+v)_{k+l-1,a^+} + (P^-u, P^-v)_{k+l-1,a^-}$$

and the corresponding norm is equivalent to the one defined in Lemma 5.3.2. For abbreviation, we set $X_A = W^{(1-l)/2}_{2,A}$ in the following. As in (5.3.2) define

$$a^k_{l-1} = a_{l-1} + (k-1/2)(i\xi)^{-1} D_x a_l \,.$$

Since $a^k_{l-1} = a'_{l-1}$ for $k = (1-l)/2$, it follows from Theorem 5.3.3 that

$$A : X_A \rightarrow W_2^{(1-l)/2} \tag{6.4.2}$$

is a Fredholm operator with index 0 if (6.4.1) is satisfied. We consider the equation

$$Au = f \;,\; f \in W_2^{(1-l)/2}. \tag{6.4.3}$$

<u>Theorem 6.4.1.</u> Assume (6.4.1) and $1/2 < d$, and suppose (6.4.2) is invertible. If $\bar{h}$ is sufficiently small, then the Galerkin equations (6.3.5) are uniquely solvable for any $f \in W_2^{(1-l)/2}$, and the approximate solutions u_Δ converge in $W_2^{(1-l)/2}$ to the exact solution u of (6.4.3) with the error bound

$$\| u-u_\Delta \|_{(1-l)/2} \leq c \min_{v \in S_d(\Delta)} \| u-v \|_{X_A} \quad (\bar{h} \rightarrow 0) \,. \tag{6.4.4}$$

<u>Proof.</u> We proceed similarly as in Theorem 6.3.1. Hypothesis (6.4.1) enables us to use Melin's inequality instead of the Gårding inequality. Let P_Δ be the orthogonal projection of $W_2^{(1-l)/2}$ onto $S_d(\Delta)$, and $P^*_\Delta \in L(W_2^{(1-l)/2})$ the adjoint of P_Δ with respect to the L_2 scalar product. Let further R_Δ be the orthogonal projection of X_A onto $S_d(\Delta)$ which

exists in view of $1/2 < d$ and the continuous embedding $W_2^{(l+1)/2} \subset X_A$.
Then equations (6.3.5) can be written

$$A_\Delta u_\Delta := P_\Delta^* A\, R_\Delta u_\Delta = P_\Delta^* f.$$

First, we show the inequality

$$\|A_\Delta u_\Delta\|_{(l-1)/2} \geq c\, \|u_\Delta\|_{(l-1)/2} \tag{6.4.5}$$

for all $u_\Delta \in S_d(\Delta)$ and sufficiently small $\bar{h}$. Arguing by contradiction, suppose there exist a sequence of meshes $\{\Delta_j\}$ and a sequence $\{u_j\}$ of elements $u_j \in S_d(\Delta_j)$ such that the mesh size of Δ_j tends to 0, $\|u_j\|_{(l-1)/2} = 1$ and $\|A_{\Delta_j} u_j\|_{(l-1)/2} \to 0$ as $j \to \infty$.
We may assume that u_j converges weakly in $W_2^{(l-1)/2}$ to some element u.
We shall prove that $u = 0$. Let Q_Δ be the orthogonal projection of $W_2^{(l+1)/2}$ onto $S_d(\Delta)$. Since Q_{Δ_j} converges strongly to the identity in $W_2^{(l+1)/2}$, we obtain

$$(A_{\Delta_j} u_j, Q_{\Delta_j} v)_0 = (u_j, A^* Q_{\Delta_j} v)_0 \to (u, A^* v)_0 = (A\, u, v)_0 \quad (j \to \infty)$$

for any $v \in W_2^{(l+1)/2}$, where $A^* \in L(W_2^{(l+1)/2}, W_2^{(l-1)/2})$ is the formal adjoint of the PDO A. Since by assumption $A_{\Delta_j} u_j \to 0$ in $W_2^{(l-1)/2}$ we have $(A\, u, v)_0 = 0$ for any $v \in W_2^{(l+1)/2}$, hence $A\, u = 0$. Since $A \in L(X_A, W_2^{(l-1)/2})$ is invertible, we get $u = 0$.

Furthermore, by Corollary 5.2.5 we deduce

$$\begin{aligned} 0 < c = c\|u_j\|^2_{(l-1)/2} &\leq \|u_j\|^2_{(l-2)/2} + \mathrm{Re}(A_{\Delta_j} u_j, u_j)_0 \\ &\leq \|u_j\|^2_{(l-2)/2} + \|A_{\Delta_j} u_j\|_{(l-1)/2} \to 0,\ j \to \infty, \end{aligned}$$

hence a contradiction which proves (6.4.5).

To verify the error estimate (6.4.4), we use the equality

$$A_\Delta(u_\Delta - R_\Delta u) = P_\Delta^*\, A(I - R_\Delta)u \tag{6.4.6}$$

which is a consequence of (6.4.3) and (6.3.5). By Theorem 6.2.3 and the Banach-Steinhaus theorem, R_Δ converges strongly to the identity in X_A as $\bar{h} \to 0$. Moreover, P_Δ^* converges strongly to the identity in $W_2^{(1-l)/2}$. Together with (6.4.5) and (6.4.6) and the continuity of $A : X_A \to W_2^{(1-l)/2}$, this implies

$$\| u_\Delta - R_\Delta u \|_{(1-l)/2} = \| A_\Delta^{-1} P_\Delta^* A(I-R_\Delta)u \|_{(1-l)/2} \leq c \, \|u-R_\Delta u\|_{X_A}$$

for all sufficiently small $\bar{h}$. Combining the last estimate with the obvious inequality

$$\|u-u_\Delta\|_{(1-l)/2} \leq \|u_\Delta - R_\Delta u\|_{(1-l)/2} + \|u-R_\Delta u\|_{X_A} ,$$

we obtain (6.4.4). □

Using Theorem 6.2.2, we now deduce asymptotic error estimates in a range of Sobolev spaces. For $k \in \mathbb{R}$, we introduce the following condition:

$\operatorname{Re} a_{l-1}^k = \operatorname{Re}\{ a_{l-1} + (k-1/2)(i\xi)^{-1} D_x a_l \} > 0$ at all points $(x,\pm 1)$, $x \in \mathbb{R}$, where $a_l(x,\pm 1)$ vanishes to first order. (C_k)

<u>Theorem 6.4.2.</u> Assume (6.4.1), $s \leq r \leq d$, $s < d-1/2$, $1/2 < d$, $(1-l)/2 \leq r$, $f \in W_2^{r-l+1}$, (C_{r-l+1}) and the invertibility of (6.4.2). For $s > (1-l)/2$ (resp. $s < (1-l)/2$), assume in addition $\Delta \in \mathcal{D}_\gamma$ for some $\gamma > 0$ (resp. (C_{s-l+1})). Then for any sufficiently small $\bar{h}$,

$$\|u-u_\Delta\|_s \leq c\|u\|_{W_{2,A}^{r-l+1}} \begin{cases} \bar{h}^{r-s} & \text{if } d \geq l-1-s , \\ \bar{h}^{d+r-l+1} & \text{if } d \leq l-1-s , \end{cases} \qquad (6.4.7)$$

where u and u_Δ are the solutions of (6.4.3) and (6.3.5), respectively.

<u>Proof.</u> First, we claim from Theorem 5.4.4, or rather its proof, that $u \in W_{2,A}^{r-l+1}$ by (C_{r-l+1}), (6.4.1) and $Au \in W_2^{r-l+1}$. Next, we extend Theorem 6.2.2 to the spaces $W_{2,A}^k$:

Let $s \leq r \leq d$, $s < d-1/2$ and $\sigma < s$. Then for any $u \in W_{2,A}^{r-l+1}$ and any partition Δ, there exists $u_\Delta \in S_d(\Delta)$ such that

$$\| u-u_\Delta \|_{W_{2,A}^{t-l+1}} \leq c(t)\bar{h}^{r-t} \| u \|_{W_{2,A}^{r-l+1}} , \; t \in [\sigma, s] . \qquad (6.4.8)$$

Indeed, by virtue of $P^\pm u \in W^r_{2,\mathbb{R}^\pm}$ and Theorem 6.2.2, we may choose elements $u_\Delta^\pm \in S_d(\Delta)$ such that

$$\| P^\pm u - u_\Delta^\pm \|_{t,\mathbb{R}^\pm} \leq c(t)\bar{h}^{r-t} \| P^\pm u \|_{r,\mathbb{R}^\pm} , \; t \in [\sigma, s] ,$$

which implies (6.4.8) with $u_\Delta = u_\Delta^+ + u_\Delta^-$.

As in the proof of Theorem 6.3.2, we now consider three cases.

1. Let $s = (l-1)/2 \le r$. Then (6.4.7) follows from (6.4.4) and (6.4.8).

2. Suppose $(l-1)/2 < s \le r$. By Theorem 6.1.3, for arbitrary $\chi \in S_d(\Delta)$,

$$\begin{aligned} \| u-u_\Delta \|_s &\le \| u-\chi \|_s + \| u_\Delta - \chi \|_s \\ &\le \| u-\chi \|_s + c\, \bar{h}^{(l-1)/2-s} \| u_\Delta - \chi \|_{(l-1)/2} . \end{aligned} \qquad (6.4.9)$$

Since the norm of the operators $A_\Delta^{-1} P_\Delta^* A \in L(X_A, W_2^{(l-1)/2})$ is uniformly bounded for all sufficiently small $\bar{h}$ (see the proof of Theorem 6.4.1), we deduce that

$$\| u_\Delta - \chi \|_{(l-1)/2} = \| A_\Delta^{-1} P_\Delta^* A(u-\chi) \|_{(l-1)/2} \le c \| u-\chi \|_{X_A} .$$

Together with (6.4.9), this implies

$$\| u-u_\Delta \|_s \le \| u-\chi \|_{W_{2,A}^{s-l+1}} + c\, \bar{h}^{(l-1)/2-s} \| u-\chi \|_{W_{2,A}^{(1-l)/2}} \qquad (6.4.10)$$

for any $\chi \in S_d(\Delta)$. Due to (6.4.8), there exists $\chi \in S_d(\Delta)$ such that the right-hand side of (6.4.10) can be estimated by $c\, \bar{h}^{r-s} \| u \|_{W_{2,A}^{r-l+1}}$, which proves (6.4.7).

3. Let $s < (l-1)/2$. By (C_{s-l+1}) and Theorem 5.3.3, $A \in L(W_{2,A}^{s-l+1}, W_2^{s-l+1})$ is a Fredholm operator with index 0. Since $A \in L(X_A, W_2^{(1-l)/2})$ is invertible and $W_2^{(1-l)/2}$ is dense in W_2^{s-l+1}, $A \in L(W_{2,A}^{s-l+1}, W_2^{s-l+1})$ is invertible, too. Therefore $A^* \in L(W_2^{l-s-1}, (W_{2,A}^{s-l+1})^*)$ is invertible, where A^* is the formal adjoint of A and $(W_{2,A}^{s-l+1})^*$ the antidual of $W_{2,A}^{s-l+1}$ with respect to the L_2 inner product. Furthermore, by (C_{s-l+1}) and Theorem 5.3.3, $A^* \in L(W_{2,A}^{-s}, W_2^{-s})$ is a Fredholm operator with index 0 since A^* satisfies condition (C_{-s}). Because of $W_{2,A}^{-s} \subset W_2^{l-s-1}$ and the invertibility of $A^* \in L(W_2^{l-s-1}, (W_{2,A}^{s-l+1})^*)$, we finally deduce that $A^* \in L(W_{2,A}^{-s}, W_2^{-s})$ is invertible.

For $g \in W_2^{-s}$, let w be the unique solution of the equation $A^* w = g$ in $W_{2,A}^{-s}$. Then we have

$$(u-u_\Delta, g)_0 = (u-u_\Delta, A^* w)_0 = (u-u_\Delta, A^*(w-v))_0$$

for any $v \in S_d(\Delta)$, which implies the estimate

$$|(u-u_\Delta, g)_0| \le c \min_{v \in S_d(\Delta)} \|w-v\|_{W_{2,A}^{(l-1)/2}} \|u-u_\Delta\|_{(l-1)/2} . \qquad (6.4.11)$$

Furthermore, (6.4.8) yields

$$\min_{v \in S_d(\Delta)} \|w-v\|_{W_{2,A}^{(l-1)/2}} \le c \begin{cases} \bar{h}^{(l-1)/2-s} \|w\|_{W_{2,A}^{-s}} & \text{if } d \ge l-1-s, \\ \bar{h}^{d-(l-1)/2} \|w\|_{W_{2,A}^{d-l+1}} & \text{if } d \le l-1-s. \end{cases} \qquad (6.4.12)$$

Since

$$\|w\|_{W_{2,A}^{-s}} \le c \|g\|_{-s} , \; \|u-u_\Delta\|_{(l-1)/2} \le c \, \bar{h}^{r-(l-1)/2} \|u\|_{W_{2,A}^{r-l+1}}$$

(cf. the first case), we obtain (6.4.7) from (6.3.13), (6.4.11) and (6.4.12). □

Remark 6.4.3. If the principal symbol of A has only zeros of order ≥ 2, then conditions (C_{r-l+1}) and (C_{s-l+1}) can be dropped in Theorem 6.4.2.

Corollary 6.4.4. Let A be elliptic and assume (6.3.3), (6.4.1), $s \le r \le d$, $1/2 < d$, $(l-1)/2 \le r$, $f \in W_2^{r-l+1}$ and, if $s > (l-1)/2$, $\Delta \in \mathcal{D}_\gamma$. Then for all sufficiently small $\bar{h}$,

$$\|u-u_\Delta\|_s \le c \|u\|_{r+1} \begin{cases} \bar{h}^{r-s} & \text{if } d \ge l-1-s , \\ \bar{h}^{d+r-l+1} & \text{if } d \le l-1-s . \end{cases} \qquad (6.4.13)$$

Proof. It follows from the ellipticity of A that $W_{2,A}^{k} = W_2^{k+l}$ for any $k \in \mathbb{R}$. Furthermore, by (6.3.3), (6.4.1), Theorem 5.1.14 and Corollary 5.1.15, $A \in L(W_2^{k+l}, W_2^{k})$ is invertible for all k. Thus for $s < d-1/2$, estimate (6.4.13) is an immediate consequence of (6.4.7). Moreover, an inspection of the proof of Theorem 6.4.2 shows that (6.4.13) is valid for $s \le d$ since estimate (6.4.10) can be replaced by

$$\|u-u_\Delta\|_s \le \|u-\chi\|_s + c \, \bar{h}^{(l-1)/2-s} \|u-\chi\|_{(l+1)/2}$$

and Theorem 6.1.1 may be used instead of (6.4.8). □

Remark 6.4.5. It follows from Theorem 6.4.2 that the maximal rate of convergence achieved by the Galerkin method under hypothesis (6.4.1) is $O(\bar{h}^{2d+1-l})$ in W_2^{l-d-1}. Thus the order of convergence is by one smaller than in the strongly elliptic case, in general.

Example 6.4.6. Consider the PDO $A = i\, S_\Gamma + c(t)V + K$, where S_Γ, V and

K are defined as in (5.5.6). A is an elliptic operator with principal symbol $i\xi/|\xi|$ and subprincipal symbol $c(t)/|\xi|$. Assume Re $c(t) > 0$ $(t \in \Gamma)$, (6.3.3) and $d \geq 1$. Then by Corollary 6.4.4 and Theorem 6.4.1, the Galerkin method (6.3.5) converges in $W_2^{-1/2}$ for any right-hand side $f \in W_2^{1/2}$, and for $s \leq r \leq d$, $-1/2 \leq r$, $f \in W_2^{r+1}$ and $\Delta \in \mathcal{D}_\gamma$ if $s > -1/2$,

$$\| u-u_\Delta \|_s \leq c \|u\|_{r+1} \begin{cases} \bar{h}^{r-s} & \text{if } d \geq -1-s , \\ \bar{h}^{d+r+1} & \text{if } d \leq -1-s . \end{cases}$$

6.4.2. Finally, we shall demonstrate by means of an example that the error estimates (6.4.13) cannot be improved, in general, if A is not strongly elliptic.

Example 6.4.7. Consider the differential operator $A = D+1$, $D = d/dx$ which has the symbol $i\xi + 1$ and the subprincipal symbol 1. Thus A is elliptic and condition (6.4.1) is satisfied. Furthermore,condition (6.3.3) is fulfilled since the equation $Au = 0$ has the general solution $u(x) = C e^{-x}$, $C \in \mathbb{C}$ on $\mathbb{R}$ which is 2π-periodic if and only if $C = 0$. Let S_1^h be the space of piecewise linear 2π-periodic splines on $[-\pi,\pi]$ subordinate to the uniform partition $\Delta = \{x_j\}_0^n$, where $x_j = -\pi + jh$ and $h = 2\pi/n$. A basis of B-splines for S_1^h is given by the "hat functions"

$$\varphi_j^{(n)}(x) = (n/2\pi)^{1/2} \begin{cases} (x-x_{j-1})/(x_j-x_{j-1}), & x \in [x_{j-1},x_j] , \\ (x_{j+1}-x)/(x_{j+1}-x_j), & x \in [x_j,x_{j+1}] , \\ 0 & \text{otherwise} , \end{cases}$$

where $j=0,\dots,n-1$, and $x_{-1} = x_{n-1}$, $x_0 = x_n$ in the definition of $\varphi_0^{(n)}$. The Galerkin operator defined by (6.3.5) has the matrix representation

$$A_h = \begin{pmatrix} a_n & b_n & 0 & \dots & \dots & 0 & c_n \\ c_n & a_n & b_n & 0 & \dots & 0 & 0 \\ \dots & & \dots & & & \dots & \\ 0 & \dots & & 0 & c_n & a_n & b_n \\ b_n & 0 & \dots & \dots & 0 & c_n & a_n \end{pmatrix}$$

with respect to the given basis, where

$$a_n = (A\varphi_1^{(n)}, \varphi_1^{(n)})_0 = 2/3, \quad b_n = (A\varphi_2^{(n)}, \varphi_1^{(n)})_0 = 1/12 + n/2,$$

$$c_n = (A\varphi_0^{(n)}, \varphi_1^{(n)})_0 = 1/12 - n/2 .$$

The matrix A_h is a circulant and possesses the eigenvalues

$$\lambda_j^{(n)} = 2/3 + (1/12+n/2)e^{2\pi ij/n} + (1/12-n/2)e^{-2\pi ij/n}$$

and a system of eigenvectors

$$u_j^{(n)} = (e^{2\pi ijk/n})_{k=0}^{n-1}, \quad A_h u_j^{(n)} = \lambda_j^{(n)} u_j^{(n)}, \quad j=0,\dots,n-1 ;$$

see Marcus and Minc [1]. Furthermore $\| \varphi_j^{(n)} \|_0^2 = 2/3$ for any j and n. It is not difficult to check that for $\varphi_h = c_0\varphi_h^{(o)}+\dots+c_{n-1}\varphi_h^{(n-1)} \in S_1^h$,

$$c^{-1} \sum_{0 \le j < n} |c_j|^2 \le \|\varphi_h\|_0^2 \le c \sum_{0 \le j < n} |c_j|^2 , \tag{6.4.14}$$

where c is independent of φ_h and n; see Prößdorf and Schmidt [1].

It follows from Theorem 6.4.1 that, for any right-hand side $f \in L_2$, Galerkin's method with splines from S_1^h for the differential equation $Au = f$ converges in L_2, while in the case of a strongly elliptic PDO of first order the method would converge in W_2^s with rate $O(h^{1-s})$ for $s \in [0,1]$. We now prove that the above Galerkin method cannot converge in W_2^s $(0 < s \le 1)$ for each right-hand side $f \in L_2$. Otherwise, by the Banach-Steinhaus theorem the estimate

$$\| A_h\varphi_h \|_0 \ge c \|\varphi_h\|_s, \quad \varphi_h \in S_1^h \tag{6.4.15}$$

would be valid. We choose the eigenvalues

$$\lambda_m^{(n)} = 2/3 + (1/6)\cos(2\pi m/2m) = 1/2$$

and the corresponding eigenvectors

$$u_m^{(n)} = ((-1)^k)_0^{n-1}$$

of A_h, where $h = 1/n$ and $n = 2m$ is even. Let

$$\varphi^{(n)} = \sum_{0 \le j < n} (-1)^j \varphi_j^{(n)}.$$

Using (6.4.14), we then obtain

$$\|A_h\varphi^{(n)}\|_0 = 2^{-1}\|\varphi^{(n)}\|_0 \le c\, n^{1/2}, \quad \|\varphi^{(n)}\|_1 \ge c\, n^{3/2}; m \in \mathbb{N} , \tag{6.4.16}$$

and by the inverse property of S_1^h (see Theorem 6.1.3),

$$\|\varphi^{(n)}\|_s \geq c\, n^{s-1} \|\varphi^{(n)}\|_1 \geq c\, n^{s+1/2}, \quad m \in \mathbb{N}$$

for $s \lesssim 1$. Together with the first inequality in (6.4.16), this contradicts (6.4.15) if $s \in (0,1]$.

6.5. Comments and references

6.1. Theorems 6.1.1 and 6.1.3 were stated in Arnold and Wendland [1]. For uniform meshes and $s \leq d$, Theorem 6.1.1 can be deduced from the results of Aubin [1, Chap. 4] and Helfrich [1] or Bramble and Scott [1]. A proof of Theorem 6.1.3 in the case $s \leq d$ was sketched in Arnold and Wendland [1]. Complete proofs of these theorems were given in Elschner and Schmidt [1]. The proof of Theorem 6.1.1 is based on an argument of Helfrich which may be extended to derive the approximation result (6.1.2) simultaneously for all $t \in (-\infty, s]$; see Elschner and Schmidt [1]. In the second part of the proof of Theorem 6.1.3 we used an argument due to Babuška and Aziz [1].

6.2. The assumption $\operatorname{Re} a(x) \geq 0$ $(x \in \mathbb{R})$ is not necessary in Theorems 6.2.2 and 6.2.3 and was only introduced to simplify the presentation. Furthermore, for $s \leq d-1$, Theorems 6.2.2 and 6.2.3 may be extended to the more general finite element spaces defined in Aubin [1]; see Elschner [11].

6.3. Theorem 6.3.1 and Theorem 6.3.2 for $s \leq d+1$ and $1/2 \leq d+1$ are special cases of the results of Stephan and Wendland [1] on Galerkin methods via finite elements for strongly elliptic PDO's on compact manifolds. In this section we applied standard techniques for Galerkin's method with finite elements (cf. e.g. Nitsche [1], Strang and Fix [1], Stephan and Wendland [1]). G. Schmidt [1] proved that the strong ellipticity condition (6.3.2) in Theorems 6.3.1 and 6.3.2 may be replaced by the following weaker assumption:

There exists $\varrho \in C^\infty$ such that the operator ϱA is strongly elliptic.

Collocation methods with splines for one-dimensional singular integral and pseudodifferential equations on closed curves have been studied by Prößdorf and Schmidt [1], Arnold and Wendland [1] and Schmidt [1].

6.4. Here we improved the results of Elschner [11] in the case of the finite element spaces $S_d(\Delta)$. In this paper, certain generalizations of Theorems 6.4.1 and 6.4.2 to more general finite elements and another proof of Theorem 6.4.1 based on a general perturbation theorem of Prößdorf and Silbermann [1, Chap. 1] for projection methods can be found.

7. APPENDIX. SUBOPTIMAL CONVERGENCE OF THE GALERKIN METHOD WITH SPLINES FOR ELLIPTIC PSEUDODIFFERENTIAL EQUATIONS

7.1. It is the aim of this appendix to survey some results of the forthcoming paper J. Elschner [On suboptimal convergence of finite element methods for pseudodifferential equations on a closed curve, Math. Nachr.]. Let Γ be a simple closed C^{∞} curve in the plane. Retaining the notation of Chapters 5 and 6, we consider a classical PDO A of order 1 on Γ with symbol $a_1(x,\xi) + a_{1-1}(x,\xi) + \dots$ and subprincipal symbol a'_{1-1}. Throughout we assume that A is elliptic (i.e. $a^{\pm}(x) := a_1(x, \pm 1) \neq 0, x \in \mathbb{R}$) and invertible as a map of W_2^s onto W_2^{s-1} for any $s \in \mathbb{R}$. Let

$$(Au_\Delta, v_\Delta)_0 = (f, v_\Delta)_0 \quad \text{for all } v_\Delta \in S_d(\Delta) \tag{7.1}$$

be the Galerkin equations for the approximate solution of the equation $Au = f$ using 2π-periodic smoothest splines of degree d subordinate to the partition Δ. Throughout, all partitions Δ are supposed to be γ-quasiuniform for some $\gamma > 0$, and $c, c_1, \dots$ denote generic constants independent of the mesh size $\bar{h}$ of Δ. Combining Theorem 6.4.1 and Corollary 6.4.4, one obtains

Theorem 7.1. Assume $1/2 < d$ and (6.4.1). Then

(i) For sufficiently small $\bar{h}$, the equations (7.1) are uniquely solvable for any $f \in W_2^{(1-1)/2}$, and the approximate solutions u_Δ converge in $W_2^{(1-1)/2}$ to the exact solution u with the error bound

$$\| u-u_\Delta \|_{(1-1)/2} = c \min_{v \in S_d(\Delta)} \| u-v \|_{(1+1)/2} .$$

(ii) If $1-1-d \leq s \leq r \leq d$, $(1-1)/2 \leq r$ and $f \in W_2^{r-1+1}$, then

$$\| u-u_\Delta \|_s \leq c\bar{h}^{r-s} \| u \|_{r+1} .$$

In case of the validity of (i) and (ii) we shall say that the Galerkin method (7.1) converges suboptimally. Note that the order of convergence is by one smaller than the quasioptimal order in the strongly elliptic case. Setting $b^{\pm}(x) = \pm \operatorname{Im} a'_{1-1}(x, \pm 1)/a^{\pm}(x)$, we have the following generalization of Theorem 7.1.

Theorem 7.2. The Galerkin method (7.1) converges suboptimally if A satisfies the following two conditions:

(V_1) For any $x_0 \in \mathbb{R}$, either $a^-(x_0)/a^+(x_0) \notin (-\infty,0)$, or $a^-(x_0)/a^+(x_0) \in (-\infty,0)$ and there is an open neighborhood U of x_0 such that Im a^-/a^+ does not change sign in U.

(V_2) If $a^-(x)/a^+(x) \in (-\infty,0)$ for all $x \in \mathbb{R}$, then

$$\int_{-\pi}^{\pi} \max(b^+(x),b^-(x))dx < 0 \quad \text{or} \quad \int_{-\pi}^{\pi} \min(b^+(x),b^-(x))dx > 0.$$

Special cases. 1. Let $A = \sum_{0 \le j \le l} a_j(x) D_x^j$ be an elliptic differential operator of order l on Γ. If l is even, then the operator $(-1)^{l/2} a_l^{-1} A$ is strongly elliptic so that Galerkin's method converges with quasi-optimal rate (cf. Schmidt [1]). Let l be odd. Then $a^-/a^+ = -1$, and (V_1) is obviously satisfied. Furthermore, it is easy to check that (V_2) is equivalent to $\int_{-\pi}^{\pi} \mathrm{Re}(a_{l-1}/a_l)dx \neq 0$.

2. Suppose the principal symbol of the PDO A is independent of x, i.e. $a^{\pm}(x) = \alpha^{\pm} \in \mathbb{C}$, and $\alpha^-/\alpha^+ \in (-\infty,0)$. Then $b^{\pm}$ in (V_2) can be replaced by $\pm$ Im $a_{l-1}(x,\pm 1)/a^{\pm}(x)$. If, in addition, $a_{l-1}(x,\pm 1) = \beta^{\pm} \in \mathbb{C}$, then ($V_2$) is equivalent to $\mathrm{Im}(\beta^+/\alpha^+)\,\mathrm{Im}(\beta^-/\alpha^-) < 0$.

7.2. Proof of Theorem 7.2. We first reduce the assertion to the case d = 1. Define the PDO $\mathcal{D}^r$, $r \in \mathbb{Z}$, by

$$\mathcal{D}^r u = (2\pi)^{-1/2}\Big\{\sum_{m \neq 0} m^r \hat{u}_m e^{imx} + \hat{u}_0\Big\},\ u \in C^\infty,$$

where $\hat{u}_m$ are the Fourier coefficients of u. $\mathcal{D}^r$ has the complete symbol ξ^r. Further $(\mathcal{D}^r u, \mathcal{D}^r v)_{s-r} = (u,v)_s$ for any $u,v \in W_2^s$, $s \in \mathbb{R}$, and $\mathcal{D}^r$ is an isomophic map of W_2^s resp. $S_d(\Delta)$ onto W_2^{s-r} resp. $S_{d-r}(\Delta)$, $d \geq r$, with the inverse $\mathcal{D}^{-r}$. Set $B = \mathcal{D}^{1-d} A \mathcal{D}^{1-d}$, $g = \mathcal{D}^{1-d} f$, $\tilde{u}_\Delta = \mathcal{D}^{1-d} u_\Delta$, $\tilde{v}_\Delta = \mathcal{D}^{1-d} v$ and $\tilde{u} = \mathcal{D}^{1-d} u$. Then the method (7.1) converges suboptimally if and only if so does the Galerkin method

$$(B\tilde{u}_\Delta, \tilde{v}_\Delta)_0 = (g, \tilde{v}_\Delta) \quad \text{for all } \tilde{v}_\Delta \in S_1(\Delta) \tag{7.2}$$

for the approximate solution of the equation $B\tilde{u} = g$. Note that B is a PDO of order $\bar{l} = l+2-2d$. Let R_Δ resp. P_Δ be the orthogonal projections of $W_2^{(\bar{l}+1)/2}$ resp. $W_2^{(\bar{l}-1)/2}$ onto $S_1(\Delta)$, and P_Δ^* the adjoint of P_Δ with respect to the L_2 scalar product. As in 6.4, the suboptimal convergence of (7.2) follows from the stability of the method, i.e. the estimate

$$\|B_\Delta u_\Delta\|_{(1-\bar{l})/2} \geq c\|u_\Delta\|_{(\bar{l}-1)/2} \quad , \; u_\Delta \in S_1(\Delta) \tag{7.3}$$

for all sufficiently small $\bar{h}$, where $B_\Delta = P_\Delta^* B R_\Delta$. Let $\varrho \in C^\infty$ such that $\varrho(x) \neq 0, x \in \mathbb{R}$. Put $\mathcal{D}_\varrho = \mathcal{D}\varrho^2 \mathcal{D}^{-2}$ and $B_{\Delta,\varrho} = P_\Delta^* \mathcal{D}_\varrho B R_\Delta$.

<u>Lemma 7.3.</u> The Galerkin method (7.2) is stable if inequality (7.3) holds with B_Δ replaced by $B_{\Delta,\varrho}$.

<u>Proof.</u> We have

$$\begin{aligned} B_\Delta &= P_\Delta^* \mathcal{D}_\varrho \mathcal{D}_{\varrho^{-1}} B R_\Delta \\ &= P_\Delta^* \mathcal{D}_{\varrho^{-1}} (I-P_\Delta^*) \mathcal{D}_\varrho B R_\Delta + \mathcal{D}_{\varrho^{-1}} B_{\Delta,\varrho} - (I-P_\Delta^*) \mathcal{D}_{\varrho^{-1}} B_{\Delta,\varrho}. \end{aligned}$$

Together with estimate (7.3) for $B_{\Delta,\varrho}$ and the inequalities

$$\|P_\Delta^* \mathcal{D}_{\varrho^{-1}} (I-P_\Delta^*)\|_{L(W_2^{(1-\bar{l})/2})} \leq c\bar{h}^{\,5/4}, \tag{7.4}$$

$$\|(I-P_\Delta^*) \mathcal{D}_{\varrho^{-1}} P_\Delta^*\|_{L(W_2^{(1-\bar{l})/2})} \leq c\bar{h}, \tag{7.5}$$

this yields

$$\begin{aligned} \|B_\Delta u_\Delta\|_{(1-\bar{l})/2} &\geq c\|u_\Delta\|_{(\bar{l}-1)/2} - c_1 \bar{h}^{5/4} \|u_\Delta\|_{(\bar{l}+1)/2} \\ &\quad - c_2 \bar{h} \|B_\Delta u_\Delta\|_{(1-\bar{l})/2}, \; u_\Delta \in S_1(\Delta), \end{aligned}$$

hence (7.3) by the inverse property of $S_1(\Delta)$. Estimate (7.4) follows by duality and the inverse property from the following lemma.

<u>Lemma 7.4.</u> Let π_Δ be the orthogonal projection of W_2^1 onto $S_1(\Delta)$. Then

$$\|(I-\pi_\Delta)\mathcal{D}^{-2}\varrho \mathcal{D}^2 u_\Delta\|_1 \leq c\bar{h}\|u_\Delta\|_{3/4} \; , \; u_\Delta \in S_1(\Delta).$$

<u>Proof of Lemma 7.4.</u> Let $Q_\Delta : W_2^1 \to S_1(\Delta)$ be the interpolation projection defined by $(Q_\Delta u)(x_k) = u(x_k)$, $k = 0,\dots,n-1$, where $x_0,\dots,x_n$ are the mesh points in $[-\pi,\pi]$. Define further

$$J u = (2\pi)^{-1}\int_{-\pi}^{\pi} u\,dx, \quad J_\Delta u = \sum_{1\leq k\leq n} u(x_k)(x_{k+1}-x_{k-1})/4\pi \; , \; u \in W_2^1.$$

Using the relation $\pi_\Delta - Q_\Delta = (J-J_\Delta)(I-\pi_\Delta)$ (cf. Arnold and Wendland [1],

Schmidt [1]), one obtains that the operator $N_\Delta = \pi_\Delta \bar{\varrho}(I-\pi_\Delta)$ can be represented as $N_\Delta u = (u,\varphi_\Delta)_1 + (u,\psi_\Delta)_1 Q_\Delta \bar{\varrho}$ with certain $\varphi_\Delta, \psi_\Delta \in W_2^1$. It may be proved that $\|\varphi_\Delta\|_1, \|\psi_\Delta\|_1 \leq c\bar{h}$. Passing to the adjoint

$$N_\Delta^* u = (I-\pi_\Delta)\mathcal{D}^{-2}\bar{\varrho}\mathcal{D}^2\pi_\Delta u = (u,1)_1\varphi_\Delta + (u,Q_\Delta\bar{\varrho})_1\,\psi_\Delta$$

of N_Δ in W_2^1, we get

$$\|N_\Delta^* u_\Delta\|_1 \leq c\bar{h}\|u_\Delta\|_{3/4} + c\bar{h}\|Q_\Delta\bar{\varrho}\|_{5/4}\|u_\Delta\|_{3/4} \leq c_1\bar{h}\|u_\Delta\|_{3/4}. \ \square$$

The proof of (7.5) relies on the estimate

$$\|(I-P_\Delta)\varrho u_\Delta\|_{(\bar{l}-1)/2} \leq c\bar{h}\|u_\Delta\|_{(\bar{l}-1)/2},\ u_\Delta \in S_1(\Delta),$$

which is a special case of corresponding results in D. N. Arnold, W. Wendland [The convergence of spline collocation for strongly elliptic equations on curves, in preparation], S. Prößdorf [Ein Lokalisierungsprinzip in der Theorie der Splineapproximationen und einige Anwendungen, Math. Nachr. 119 (1984)] or G. Schmidt [On ε-collocation for pseudodifferential equations on closed curves, Math. Nachr., to appear]. □

Now we condider the PDO $B_\varrho = \mathcal{D}_\varrho B$ which has the principal symbol $\varrho\xi^{2-2d}a_l$ and the subprincipal symbol

$$\xi^{2-2d}\{\varrho a'_{l-1} - (2i)^{-1}D_\xi a_l D_x\varrho + (i\xi)^{-1}[2a_l D_x\varrho + (1-d)\varrho D_x a_l]\}.$$

It can be proved that A satisfies the conditions (V_1) and (V_2) if and only if there exists $\varrho \in C^\infty, \varrho(x) \neq 0$ $(x \in \mathbb{R})$, such that B_ϱ satisfies (6.4.1). By Theorem 7.1 and Lemma 7.3 we then obtain Theorem 7.2. □

It can be shown that conditions (V_1) and (V_2) are even necessary for the suboptimal convergence of (7.1) if the terms a_l and a_{l-1} of the symbol do not depend on x. It is not known whether (V_1) and (V_2) are necessary in the general case. Combining the methods of this appendix with those of Arnold and Wendland [1] and Schmidt [1] one can derive results on the suboptimal convergence of collocation methods.

REFERENCES

Agranovič, M.S. (Агранович, М.С.)

[1] Спектральные свойства эллиптических псевдодифференциальных операторов на замкнутой кривой. Функц. анализ 13 (1974) 4, 54-56.

Agranovič, M.S., and M.I. Višik (Агранович, М.С., и М.И. Вишик)

[1] Эллиптические задачи с параметром и параболические задачи общего вида. УМН 19 (1964) 3, 53-161.

Alinhac, S., and M.S. Baouendi

[1] Uniqueness for the characteristic Cauchy problem and strong unique continuation for higher order partial differential inequalities. Amer. J. Math. 102 (1980) 197-217.

Arnold, D.N., and W. Wendland

[1] On the asymptotic convergence of collocation methods. Math. Comput. 41 (1983).

Aubin, J.P.

[1] Approximation of elliptic boundary value problems. Wiley Interscience, New York 1972.

Babuška, I., and A.K. Aziz

[1] Survey lectures on the mathematical foundation of the finite element method. In: The mathematical foundation of the finite element method with applications to partial differential equations (edited by A.K. Aziz). Academic Press, New York 1972.

Bagirov, L.A., and V.A. Kondratiev (Багиров, Л.А., и В.А. Кондратьев)

[1] Об эллиптических уравнениях в R^n. Дифференц. уравнения 11 (1975) 498-504.

Baouendi, M.S., and C. Goulaouic

[1] Cauchy problems with characteristic initial hypersurface. Comm. Pure Appl. Math. 26 (1973) 455-475.

Baouendi, M.S., C. Goulaouic and J.L. Lipkin

[1] On the operator $\Delta r^2 + \mu(\partial/\partial r)r + \lambda$. J.Diff.Eq. 15 (1974) 195-209.

Baouendi, M.S., and J. Sjöstrand

[1] Nonhypoellipticité d'opérateurs elliptiques singuliers. Séminaire Goulaouic-Schwartz 1975-1976, Exposé XXIV.

[2] Regularité analytique pour des opérateurs elliptiques singuliers en un point, Ark. Mat. 14 (1976) 9-33.

Beals, R.

[1] A general calculus of pseudodifferential operators. Duke Math. J. 42 (1975) 1-42.

Blekher, P.M. (Блехер, П.М.)

[1] Об операторах, зависящих мероморфно от параметра. Вестник Моск. ун-та 24 (1969) 5, 30-36.

Bolley, P., and J. Camus

[1] Sur une classe d'opérateurs elliptiques et dégénérés a une varible. J. Math. pures appl. 51 (1972) 429-463.

Bolley, P., J. Camus and B. Helffer

[1] Sur une classe d'opérateurs partiellement hypoelliptiques. J. Math. pures appl. 55 (1976) 131-171.

[2] Hypoellipticité partielle pour des opérateurs dégénérés non-Fuchsiens. Comm. Partial Diff. Eq. 2 (1977) 1-30.

de Boor, C.

[1] A practical guide to splines. Springer-Verlag, New York 1978.

Bourbaki, N.

[1] Algèbre. Chap. IV-V. Hermann, Paris 1950.

Boutet de Monvel, L.

[1] Boundary problems for pseudo-differential operators. Acta Math. 126 (1971) 11-51.

Bramble, J.H., and R. Scott

[1] Simultaneous approximation in scales of Banach spaces. Math. Comput. 32 (1978) 947-954.

Coddington, E.A., and N. Levinson
[1] Theory of ordinary differential equations. Mc Graw-Hill, New York 1955.

Cope, F.T.
[1] Formal solutions of irregular linear differential equations I. Amer. J. Math. 56 (1934) 411–437.

Deligne, P.
[1] Equations différentielles a points singuliers reguliers. Lecture Notes Math. 163, Springer-Verlag, Berlin 1970.

Elschner, J.
[1] Über entartete gewöhnliche Differentialgleichungen. Math. Nachr. 68 (1975) 183–199.
[2] Über die normale Auflösbarkeit von entarteten gewöhnlichen Differentialoperatoren in Räumen differenzierbarer Funktionen. Math. Nachr. 81 (1978) 169–193.
[3] Über selbstadjungierte singuläre gewöhnliche Differentialoperatoren und ein Problem von H. Triebel. Math. Nachr. 93 (1979) 53–65.
[4] Über entartete gewöhnliche Differentialoperatoren und eine Klasse entarteter singulärer Integrodifferentialgleichungen. Math. Nachr. 94 (1980) 117–142.
[5] Über die normale Auflösbarkeit von entarteten gewöhnlichen Differentialoperatoren auf einem unendlichen Intervall. Math. Nachr. 98 (1980) 165–181.
[6] Über entartete Randwertprobleme für elliptische Differentialgleichungen zweiter Ordnung in der Ebene. ZIMM-Report R-04/80, Akademie der Wiss. der DDR, Inst. f. Mathem., Berlin 1980.
[7] On degenerate boundary value problems for elliptic differential operators of second order in the plane and the index of degenerate pseudo-differential operators on a closed curve. Math. Nachr. 102 (1981) 277–292.
[8] On a class of degenerate singular integro-differential equations II. Math. Nachr. 103 (1981) 255–281.
[9] On the index of degenerate pseudodifferential operators on a closed curve. Math. Nachr. 109 (1982) 79–91.
[10] On the index of hypoelliptic one-dimensional pseudodifferential operators. In: Seminar Analysis 1981/82, Akademie der Wiss. der DDR, Inst. f. Mathem., Berlin 1982.
[11] A Galerkin method with finite elements for degenerate one-dimensional pseudodifferential equations. Math. Nachr. 111 (1983) 111–126.

Elschner, J., and M. Lorenz
[1] On a class of elliptic differential operators degenerating at one point. In: Proceedings of the Banach semester "Partial Diff. Eq." 1978, PWN, Warszawa 1983.
[2] Über den Operator $\sum_{|\alpha|=|\beta|\le m} a_{\alpha\beta} x^{\alpha} D^{\beta}$. Wiss. Zeitschrift TH Karl-Marx-Stadt 21 (1979) 529–534.
[3] On the normal solvability of the operator $\Delta r^2+\mu(\partial/\partial r)r+\lambda$. J. Diff. Eq. 36 (1980) 408–424.
[4] An unsolvable hypoelliptic differential operator degenerating at one point. Wiss. Zeitschrift TH Karl-Marx-Stadt 23 (1981) 369–373.

Elschner, J., and G. Schmidt
[1] On spline interpolation in periodic Sobolev spaces. Preprint P-MATH-01/83, Akademie der Wiss. der DDR, Inst. f. Mathem., Berlin 1983.

Elschner, J., and B. Silbermann
[1] Eine Klasse entarteter gewöhnlicher Differentialgleichungen und das Kollokationsverfahren zu ihrer Lösung. Czech. Math. J. 29 (1979) 551–563.

[2] Entartete gewöhnliche Differentialoperatoren und einige ihrer Verallgemeinerungen I. Wissenschaftliche Schriftenreihe der TH Karl-Marx-Stadt 3 (1980).

Eskin, G.I. (Эскин, Г.И.)
[1] Краевые задачи для эллиптических псевдодифференциальных уравнений. Наука, Москва 1973.

Gantmacher, F.R.
[1] Matrizenrechnung II. DVW, Berlin 1959.

Gelfand, I.M., N.J. Vilenkin
[1] Verallgemeinerte Funktionen IV. DVW, Berlin 1964.

Gluško, V.P. (Глушко, В.П.)
[1] Вырождающиеся линейные дифференциальные уравнения I-IV. Дифференц. уравнения 4 (1968) 1584-1597, 4 (1968) 1956-1966, 5 (1969) 443-455, 5 (1969) 599-610.

[2] Линейные вырождающиеся дифференциальные уравнения. Воронеж 1972.

Goldberg, S.
[1] Unbounded linear operators. Mc Graw-Hill, New York 1966.

Helffer, B., and Y. Kannai
[1] Determining factors and hypoellipticity of ordinary differential operators with "double characteristics". Asterisque 2/3 (1973) 197-216.

Helffer, B., and L. Rodino
[1] Opérateurs différentiels ordinaires intervenant dans l'etude de l'hypoellipticité. Bolletino U.M.I. 14 B (1977) 491-522.

Helfrich, H.P.
[1] Simultaneous approximation in negative norms of arbitrary order. R.A.I.R.O. Numer. Anal. 15 (1981) 231-235.

Hoffman, K.
[1] Banach spaces of analytic functions. Prentice Hall, Englewood Cliffs 1962.

de Hoog, F.R., and R. Weiss
[1] Difference methods for singular boundary value problems of ordinary differential equations. SIAM J. Numer. Anal. 13 (1976) 106-134.

[2] The numerical solution of boundary value problems with an essential singularity. SIAM J. Numer. Anal. 16 (1979) 637-669.

[3] On the boundary problem for systems of ordinary differential equations with a singularity of the second kind. SIAM J. Math. Anal. 11 (1980) 41-60.

Hörmander, L.
[1] On the division of distributions by polynomials. Ark. Mat. 3 (1958) 555-568.

[2] Linear partial differential operators. Springer-Verlag, Berlin 1963.

[3] Pseudo-differential operators. Comm. Pure Appl. Math. 18 (1965) 501-517.

[4] Pseudo-differential operators and hypoelliptic equations. Amer. Math. Soc. Symp. Pure Math. 10 (1966) 138-183.

[5] The Weyl calculus of pseudo-differential operators. Comm. Pure Appl. Math. 32 (1979) 360-444.

Ince, E.L.
[1] Ordinary differential equations. London 1927.

Kannai, Y.
[1] An unsolvable hypoelliptic operator. Israel J. Math. 9 (1971) 306-315.

[2] Hypoelliptic ordinary differential operators. Israel J. Math. 13 (1972) 106-134.

[3] Hypoellipticity of certain degenerate elliptic boundary value problems. Transactions Amer. Math. Soc. 217 (1976) 311-328.

Kohn, J.J., and L. Nirenberg
[1] On the algebra of pseudo-differential operators. Comm. Pure Appl. Math. 18 (1965) 269-3o5.

Komatsu, H.
[1] On the index of ordinary differential operators. J. Fac. Sci. Univ. Tokyo, Sec. I A, 18 (1971) 379-398.

Kondratiev, V.A. (Кондратьев, В.А.)
[1] Краевые задачи для эллиптических уравнений в областях с коническими или угловыми точками. Труды Моск. Матем. общ-ва I6 (I967) 209-292.

Korobejnik, Ju.F. (Коробейник, Ю.Ф.)
[1] Нормальная разрешимость линейных дифференциальных уравнений в комплексной плоскости. Изв. АН СССР, Сер. Матем. 36 (I972) 450-47I.

Krasnoselski, M.A., et al.
[1] Näherungsverfahren zur Lösung von Operatorgleichungen. Akademie-Verlag, Berlin 1973.

Kuznecov, A.N. (Кузнецов, А.Н.)
[1] Дифференцируемые решения вырождающихся систем обыкновенных уравнений. Функц. анализ 6 (I972) 2, 4I-52.

Lewis, J.E., and C. Parenti
[1] Abstract singular parabolic equations. Comm. Partial Diff. Eq. 7 (1982) 279-324.

Lions, J.L., and E. Magenes
[1] Problèmes aux limites nonhomogenes et applications I. Dunod, Paris 1968.

Lomov, S.A. (Ломов, С.А.)
[1] Обобщение теоремы Фукса на неаналитический случай. Матем. сб. 65 (I964) 498-5II.

Lorenz, M.
[1] Unsolvable hypoelliptic differential operators with a totally characteristic point. Math. Nachr. 114 (1983).

[2] An elliptic differential operator degenerating at one point which is hypoelliptic but not locally solvable. Math. Nachr. (to appear).

Malgrange, B.
[1] Remarques sur les points singuliers des equations différentielles. C.R. Acad. Sc. Paris A 273 (1971) 1136-1137.

[2] Sur les points singuliers des equations différentielles. Enseign. Mathem. 2o (1974) 147-176.

Marcus, M., and H. Minc
[1] A survey of matrix theory and matrix inequalities. Allyn and Bacon, Boston 1964.

Melin, A.
[1] Lower bounds for pseudo-differential operators. Ark. Mat. 9 (1971) 117-14o.

Michlin, S.G.
[1] Partielle Differentialgleichungen in der mathematischen Physik. Akademie-Verlag, Berlin 1978.

Michlin, S.G., and S. Prößdorf
[1] Singuläre Integraloperatoren. Akademie-Verlag, Berlin 198o.

Miranda, G.
[1] Equazioni alle derivate parziali di tipo ellittico. Springer-Verlag, Berlin 1955.

Müller-Pfeiffer, E.
[1] Spektraleigenschaften singulärer gewöhnlicher Differentialoperatoren. Teubner-Verlag, Leipzig 1977.

Muschelischwili, N.I.
[1] Singuläre Integralgleichungen. Akademie-Verlag, Berlin 1965.

Narasimhan, R.
[1] Analysis on real and complex manifolds. Masson and Cie., Paris 1968.

Natterer, F.
[1] A generalized spline method for singular boundary value problems of ordinary differential equations. Lin. Alg. Appl. 7 (1973) 189-216.
[2] Das Randwertproblem für eine Klasse singulärer gewöhnlicher Differentialausdrücke. Aeq. Math. 12 (1975) 207-228.

Neumark, M.A.
[1] Lineare Differentialoperatoren. Akademie-Verlag, Berlin 1960.

Nitsche, J.
[1] Über ein Variationsprinzip zur Lösung von Dirichlet-Problemen bei Verwendung von Teilräumen, die keinen Randbedingungen unterworfen sind. Abh. d. Hamb. Math. Sem. 36 (1971) 9-15.

Prevosto, D., and J. Rolland
[1] Théorème d'indice et regularité pour une classe d'opérateurs elliptiques et dégénérés. C.R. Acad. Sc. Paris A 279 (1974) 873-876.

Prößdorf, S.
[1] Einige Klassen singulärer Gleichungen. Akademie-Verlag, Berlin 1974 (Engl. transl.: North Holland, Amsterdam 1978).

Prößdorf, S., and G. Schmidt
[1] A finite element collocation method for singular integral equations. Math. Nachr. 100 (1981) 33-60.

Prößdorf, S., and B. Silbermann
[1] Projektionsverfahren und die näherungsweise Lösung singulärer Gleichungen. Teubner-Verlag, Leipzig 1977.

Przeworska-Rolewicz, D., and S. Rolewicz
[1] Equations in linear spaces. PWN, Warszawa 1968.

Rempel, S., and B.-W. Schulze
[1] Index theory for elliptic boundary problems. Akademie-Verlag, Berlin 1983.

Robertson, A.P., and W.J. Robertson
[1] Topological vector spaces. Cambridge University Press. London 1964.

Seeley, R.T.
[1] Integral equations depending analytically on a parameter. Indag. Math. 24 (1962) 434-442.

Schmidt, G.
[1] The convergence of Galerkin and collocation methods with splines for pseudodifferential equation on closed curves. Zeitschrift f. Anal. und ihre Anwend. 3 (1984) 5.

Stein, E.M.
[1] Singular integrals and differentiability properties of functions. Princeton University Press, Princeton 1970.

Stein, E.M., and G. Weiss
[1] Introduction to Fourier analysis on Euklidean spaces. Princeton University Press, Princeton 1971.

Stephan, E., and W. Wendland
[1] Remarks to Galerkin and least squares methods with finite elements for general elliptic problems. In: Lecture Notes Math. 564, Springer-Verlag, Berlin 1976.

Sternberg, W.
[1] Über die asymptotische Integration von Differentialgleichungen. Math. Ann. 81 (1920) 119-186.

Strang, G., and G. Fix
[1] An analysis of the finite element method. Prentice Hall, Englewood Cliffs 1973.

Šubin, M.A. (Шубин, М.А.)
[1] Псевдодифференциальные операторы и спектральная теория. Наука, Москва 1978.

Svensson, S.L.
[1] Singular differential operators and distributions. Israel J. Math. 38 (1981) 131-153.

Taira, K.
[1] Sur le problème de la derivée oblique II. Ark. Mat. 17 (1979) 177-191.

Taylor, M.
[1] Pseudodifferential operators. Princeton University Press, Princeton 1981.

Tovmasjan, N.E. (Товмасян, Н.Е.)
[1] К теории сингулярных интегральных уравнений, Дифференц. Уравнения 3 (1967) 69-80.

Treves, F.
[1] Introduction to pseudodifferential and Fourier integral operators I. Plenum Press, New York 1980.

Triebel, H.
[1] Allgemeine Legendresche Differentialoperatoren I. J. Funct. Anal. 6 (1970) 1-25.

[2] Interpolation theory for spaces of Besov type. Elliptic differential operators. In: Theory of nonlinear operators, Proceedings Summer School Babylon (CSSR) 1971, Prag 1973.

[3] Höhere Analysis. DVW, Berlin 1972.

[4] Interpolation theory, function spaces, differential operators. DVW, Berlin 1978.

Vekua, I.N.
[1] Verallgemeinerte analytische Funktionen. Akademie-Verlag, Berlin 1963.

Višik, M.I., and V.V. Grušin (Вишик, М.И., и В.В. Грушин)
[1] Об одном классе вырождающихся эллиптических уравнений высших порядков. Матем. сб. 79 (1969) 3-36.

Wasow, W.
[1] Asymptotic expansions for ordinary differential equations. Interscience, New York 1965.

Widder, D.V.
[1] The Laplace transform. Princeton University Press, Princeton 1941.

LIST OF BASIC SYMBOLS

Spaces

Operators

Other symbols

SUBJECT INDEX